Principles and Applications of Polarization-Division Interferometry

Principles and Applications of Polarization-Division Interferometry

Edited by

Prasad L. Polavarapu
Department of Chemistry, Vanderbilt University, Nashville, TN, USA

JOHN WILEY & SONS

Chichester · New York · Weinheim · Brisbane · Singapore · Toronto

National 01243 779777
International (+44) 1243 779777
e-mail (for orders and customer service enquiries): cs-books@wiley.co.uk
Visit our Home Page on http://www.wiley.co.uk
or http://www.wiley.com

Other Wiley Editorial Offices

John Wiley & Sons, Inc., 605 Third Avenue,
New York, NY 10158-0012, USA

WILEY-VCH Verlag GmbH, Papelallee 3,
D-69469 Weinheim, Germany

Jacaranda Wiley Ltd, 33 Park Road, Milton,
Queensland 4064, Australia

John Wiley & Sons (Asia) Pte Ltd, Clementi Loop #02-01,
Jin Xing Distripark, Singapore 129809

John Wiley & Sons (Canada) Ltd, 22 Worcester Road,
Rexdale, Ontario M9W 1L1, Canada

Library of Congress Cataloging-in-Publication Data

Principles and applications of polarization division interferometry /
edited by Prasad L. Polavarapu.
p. cm.
Includes bibliographical references and index.
ISBN 0-471-97420-X (alk. paper)
1. Polarization interferometers. 2. Infrared spectroscopy.
I. Polavarapu, Prasad L.
QC411.P75 1997
543'.08583—dc21 97-17447

British Library Cataloguing in Publication Data

A catalogue record for this book is available from the British Library

ISBN 0 471 97420 X

Typeset in 10/12pt Times by Mathematical Composition Setters Ltd, Salisbury
Printed and bound in Great Britain by Biddles Ltd, Guildford. Surrey
This book is printed on acid-free paper responsibly manufactured from sustainable forestation, for which at least two trees are planted for each one used for paper production

Contents

Contributors

Lawrence B. Coleman *Department of Physics, University of California at Davis, Davis, CA 95616, USA*

Mikihiko Ikezawa *Research Institute for Scientific Measurements, Tohoku University, Aoba-ku, Sendai 980, Japan*

Derek H. Martin *Physics Department, Queen Mary and Westfield College, University of London, London E1 4NS, UK*

John C. Mather *Code 680, NASA Goddard Space Flight Center, Greenbelt, MD 20771, USA*

Takao Nanba *Department of Physics, Faculty of Science, Kobe University, Kobe 657, Japan*

Prasad L. Polavarapu *Department of Chemistry, Vanderbilt University, Nashville, TN 37235, USA*

Preface

Most of the commercial Fourier transform spectrometers [1] are based on a Michelson interferometer [2] where the input light is divided, by an amplitude-division beamsplitter, into two equal-amplitude components. For this reason, Michelson interferometers are also referred to as amplitude-division (or amplitude-dividing [3]) interferometers (ADIs). Polarization-division (or polarization-dividing) interferometers (PDIs) are another class of instruments which offer definite advantages over ADIs in certain applications. An appropriate polarization-division beamsplitter forms the central component in these interferometers. Different types of PDIs are known, but those appropriate for a broad spectral range are of interest here. One such interferometer is based on the division of polarized light by a Wollaston prism and the principles of this instrumentation are well documented [4]. The approach suggested by Martin and Puplett [5] forms the basis for another type of PDI. Here the input light is linearly polarized and divided, by a polarization-division beamsplitter made of wire grids, into two components with orthogonal linear polarizations. These two components are brought back to the beamsplitter using two roof-top mirrors, which rotate the linear polarizations by 90°. With the use of a linear polarization analyzer, placed in the output beam, the polarization modulation resulting from the mirror travel can be converted to intensity modulation and the instrument now functions as an interferometer, referred to as a Martin–Puplett interferometer (MPI). The principles of this interferometry and its applications to spectroscopic and radiometric measurements are presented by Professor Martin in Chapter 1. The importance of this interferometry can be discerned from its applications in atmospheric and space sciences, in the far-infrared and microwave regions. One such application has been the detection and analysis of cosmic radiation appearing in the far-infrared region, to verify the Big Bang theory of the origin of Universe. The details of these measurements and the current status of this application are described by Dr Mather in Chapter 2. Owing to the weak emission from black-body sources in the far-infrared region, spectroscopic applications often desire brighter sources than are normally available. In Chapter 3, Professors Nanba and Ikezawa describe the adaption of synchrotron radiation as the light source for MPI and present the spectroscopic applications in the far-infrared region. When the moving mirror in the interferometer is advanced in discrete steps, some kind of external modulation is needed for efficient signal detection. This is normally achieved in the far-infrared region with either a light chopper or a rotating analyzer. Additional advantages can be realized when the input and output polarizers are simultaneously rotated, at different frequencies. The theoretical details and practical applications of this approach are described by Professor Coleman in Chapter 4.

Despite the importance of the applications reported, the advantages of this interferometry have not been recognized by the broader scientific community. The reason for this gap might be twofold. Although a majority of the scientific applications are commonly undertaken in the mid-infrared region, all of the early applications of MPI were confined to the far-infrared region. Second, the applications of this interferometry in the chemical sciences were not widely recognized. When an MPI is configured without an output analyzer [6], the measured detector signal is proportional to the difference between polarized absorbances (or reflectances) of a sample, and a direct measurement of such difference signals provides additional advantages, and new opportunities, for probing molecular polarization properties. A realization of these, and other, opportunities provided the impetus for new research activity using wire-grid beamsplitter based PDIs. It is now established that a wire-grid beamsplitter-based PDI can be used in the mid- and far-infrared regions for a variety of spectroscopic measurements. The principles and applications of these measurements are presented in Chapter 5.

The chapters presented in this book summarize the principles, importance and advantages of wire-grid beamsplitter based PDIs, all centered around the design of Martin and Puplett. The applications presented range from establishing a support for the Big Bang theory for the origin of Universe to material characterization; in addition, opportunities for new applications are presented. It is hoped that the contents presented here will provide a valuable insight into polarization-division interferometry and help advance the role and applications of this interferometry in emerging areas of science.

Prasad L. Polavarapu
Nashville, TN
March 1997

REFERENCES

1. P.R. Griffiths and J.A. de Haseth, *Fourier Transform Infrared Spectrometry*, Wiley, New York (1988).

2. A.A. Michelson, *Light Waves and Their Uses*, University of Chicago Press, Chicago (1907).
3. K.J. Gasvik, *Optical Metrology*, Wiley, New York (1987).
4. M. Fracon and S. Mallick, *Polarization Interferometers*, Wiley-Interscience, New York (1971).
5. D.H. Martin and E. Puplett, *Infrared Phys.*, **10**, 105 (1969).
6. M.J. Dignam and M.D. Baker, *Appl. Spectrosc.*, **35**, 186 (1981).

1

The Principles of Polarization-Division Interferometric Spectrometry

DEREK H. MARTIN

Physics Department, Queen Mary and Westfield College, University of London, London E1 4NS, UK

CONTENTS

Principles and Applications of Polarization-Division Interferometry
Edited by P. L. Polavarapu

1 INTRODUCTION

1.1 TWO-BEAM INTERFEROMETRIC SPECTRORADIOMETRY AND SPECTROSCOPY

There have been many applications of two-beam polarization-division, or Martin–Puplett, interferometers in wide-band spectrometry since the introduction of the technique in 1970 [1]. These include a number of major scientific studies in which the special merits of this spectrometric method have been of crucial importance. Such applications are the subjects of the other chapters of this book. The principles of the method are the concern here.

The power of two-beam interferometry for wide-band spectrometric measurements had been demonstrated before the introduction of polarization-division interferometers. Amplitude-division, or Michelson, interferometers had been used in spectroscopic studies of weak signal beams, in particular in studies of solar and planetary infrared sources and of atmospheric and laboratory sources of far-infrared radiation. Michelson's own measurements of the fine structures of lines in atomic spectra in the visible range, 100 years ago, were the first applications of two-beam interferometry for spectrometric purposes but the use of such interferometers for wide-band spectrometry had to wait on technical developments of the 1960s, including wide-band detectors with electrical outputs linear over large dynamic ranges of incident signal power, high dynamic range analogue-to-digital converters, digital data storage, digital computers and efficient algorithms for the computation of discrete Fourier transforms. Once these were available, the special advantages of two-beam interferometric spectrometry over the then prevalent dispersive methods of spectrometry (based on diffraction gratings or prisms) were impressively realized. The first of these advantages is the larger angular size of source that is allowable, for a given spectral resolution, in the interferometric method, i.e. a larger throughput and therefore a larger signal power. The second is a reduced consequence of detector noise (the intrinsic fluctuations in the output from the detector) in the interferometric method; all the signal power is incident on the detector throughout the measurement time, rather than that falling within a single resolution interval, and as a consequence of this 'frequency multiplexing,' the noise in each resolution interval of a measured spectrum is less. These throughput and multiplexing advantages [2] maximize the signal-to-noise ratio in a spectrum and make the interferometric method especially suited to studies involving low signal beam powers. Many amplitude-division Michelson interferometric spetrometers are now in use in the far-infrared, infrared and visible regions [3].

Polarization-division or Martin–Puplett, two-beam interferometers [1, 4] have the same throughput and multiplexing advantages as Michelson interferometers and they bring a number of additional advantages. Highly efficient beam division is maintained over wider continuous ranges of frequency. All four ports of the interferometer are isolated and accessible, and the optical symmetry of the interferometer can therefore be used, in balanced modes of operation, to suppress noise arising from internal dimensional instabilities and from fluctuations in total source power, and to avoid radiometric contamination attributable to temperature differences between the detector and the interferometer. Calibration sources can be incorporated in the second input port. Moreover, the method will determine directly the spectrum of the polarization state of a wide-band thermal signal.

The distribution of signal power in the interference pattern produced by an interferometer will depend on the spatial spectrum of the incident beam (the distribution of propagation directions within it) as well as on its temporal spectrum (the frequency distribution). The two could not generally be separated in an analysis of the pattern. When spatial spectra are the subjects of interest, as in optical metrological applications, it is necessary to band-limit strongly the temporal spectrum of the signal beam, i.e. the signal beam must be quasi-monochromatic. Correspondingly, when the measurement of a temporal spectrum of a signal beam is required over a wide spectral band, as in spectrometry, the spatial spectrum of the beam must be band-limited, i.e. the beam must be well collimated. There are many polarisation-division interferometers in use for metrological purposes but they differ from the Martin–Puplett interferometers which are the subject of this book because they incorporate narrow-banded multi-layer interference structures as polarisation beam dividers. The instruments of concern here are for wide-band spectrometric purposes and incorporate wide-band beam dividers and polarisation rotators.

There are two distinct objectives in spectrometry. In the first, 'spectroradiometry,' the purpose is the determination of the spectrum of a wide-band signal beam from a (perhaps remote) source of interest. In the second, 'spectroscopy,' the purpose is the measurement of the reflectance or transmitance of a sample of a material of interest, or of an optical component, as a function of frequency over a wide frequency range. The instruments of concern here can serve either purpose. A wide-band signal beam is involved in both; in spectroradiometry the beam is the object of interest but in spectroscopy it is a means to an end.

The wide-bandedness of the signal beam in either case derives from the fact that the electric field amplitude at a point in the beam will be fluctuating in a random manner in time. Such beams are radiated from matter (gases, solids, plasmas) in which the motions of ions and electrons are in constant thermal agitation, with rapid random exchanges of energy between them, and with some leakage of electromagnetic quanta from the matter. The leaking quanta constitute the thermal radiation. The random fluctuations of the electric field at a point in the thermal radiation beam is a consequence of the irregular contributions arising from the random motions of countless charged ions and electrons in the source material.

In a spectroradiometric study, it is aspects of the structures and motions of the atoms, molecules, electrons and ions in the radiating material, as inferred from the measured spectra, that are of interest. In such cases the temperature of the radiating material may not be high and the radiated power from cool matter is weak and its spectrum is limited to longer wavelengths (its power spectrum will lie everywhere below the black body curve for the temperature in question). For that reason, interferometric spectrometry has been essential in such studies.

Spectroscopic studies require that the signal beam should, with as much power as possible, cover the range of frequencies over which the reflective or transmissive properties of the sample under study are to be determined. That points to a high temperature for the source, e.g. gas in a high-pressure discharge tube or an electrically heated ceramic rod.

Signal beams in which the electric field at each point fluctuates randomly in time are referred to as 'thermal' or 'noise' beams. To understand the function of an interferometric spectrometer, it is necessary to be able to describe and quantify the properties of such beams and the way those properties change on propagation through a linear system. That will be examined in the following section.

1.2 THE CHARACTERISTIC PROPERTIES OF WIDE-BAND THERMAL SIGNALS

In a thermal or noise signal beam that has been collimated and linearly polarized the electric field, at each point in the beam, is transverse to the axis of propagation and fluctuates in time in a random manner. This time-dependent field at a point in the beam will be denoted $E(t)$. The properties that characterize the beam are statistical properties of $E(t)$. Those of importance in spectrometry are examined below. Full discussions of such matters can be found in standard texts, such as Ref. [5], including proofs of relationships that are simply stated below. It will be seen later that an interferometer essentially measures the degree of correlation between the fluctuating fields in two thermal beams one of which has been delayed in time relative to the other. Quantitative measures of correlation are therefore included below, as are the relationships between the correlation properties of signal beams and their spectra. The angle brackets $\langle\ \rangle$ denote the time-averaged values of the quantity enclosed.

Mean value, $\langle E(t)\rangle$

The E-field lies in the direction of polarization but fluctuates in magnitude and sense, with zero mean-value: $\langle E(t)\rangle = 0$.

Mean-square value, or signal power, $\langle E^2(t)\rangle$

The mean-square of the E-field, $\langle E^2(t)\rangle$, is of definite value for a given beam and is customarily called the signal power of the beam. [To obtain the true power (watts) carried by the beam, this signal power must be multiplied by the cross-sectional area of the beam and by the characteristic impedance of the medium of propagation, usually free-space].

Probability distribution

The probability that the value of $E(t)$, taken at a randomly selected value of t, lies between E and $(E + \mathrm{d}E)$ decreases exponentially with E^2, with a standard deviation equal to $\langle E^2(t)\rangle$.

The auto-correlation function, $A(\tau)$

$E(t)$ is a continuous function of time, varying at a finite rate at all instants in time. The values of $E(t)$ at two sufficiently closely spaced instants in time would therefore not differ greatly, i.e. the values would be highly correlated in the sense that the probability distribution for the value of $E(t)$ at the later instant will be strongly peaked at the value taken at the earlier instant. The values of $E(t)$ at two sufficiently widely spaced instants in time, however, would be uncorrelated, i.e. the probability that $E(t)$ at the later time would take any particular value would be given by the single-instant distribution referred to above, independently of the value taken at the earlier time. The degree of correlation between the values of $E(t)$ at instants separated by a time interval τ, changes with increasing τ, and the way in which it varies, from complete correlation at $\tau = 0$ to zero correlation as $\tau \longrightarrow \infty$, for any signal beam is an important property of that beam. This is the beam's auto-correlation function, $A(\tau)$ say, defined as the average over time t of the product $\{E(t+\tau)E(t)\}$:

$$A(\tau) \equiv \langle E(t+\tau)E(t)\rangle$$

Given the averaging over time, it will be clear that $\langle E(t+\tau)E(t)\rangle \equiv \langle E(t+T+\tau)E(t+T)\rangle$ for any value of T. Collimated beams are being considered here, so the function $E(t)$ denoting the time dependence of the E-field at a fixed point in the beam will also denote the distance dependence of the field along the propagation direction at a fixed instant in time, if t is read as d/c, where d is distance along the beam and c is the velocity of light. The autocorrelation function (ACF) is clearly the same at all points in a freely propagating collimated beam.

The ACF of a linearly polarized collimated beam is thus a statement of the statistical properties of the beam expressed *in the time domain.* It is symmetrical about $\tau = 0$ and has its absolute maximum at $\tau = 0$; as $|\tau|$ increases from zero $A(\tau)$ will decrease and eventually approach zero asymptotically, but it may, in between, become negative over some ranges of $|\tau|$. An illustrative example of an autocorrelation function for a collimated thermal signal beam is shown in Figure 1 (a).

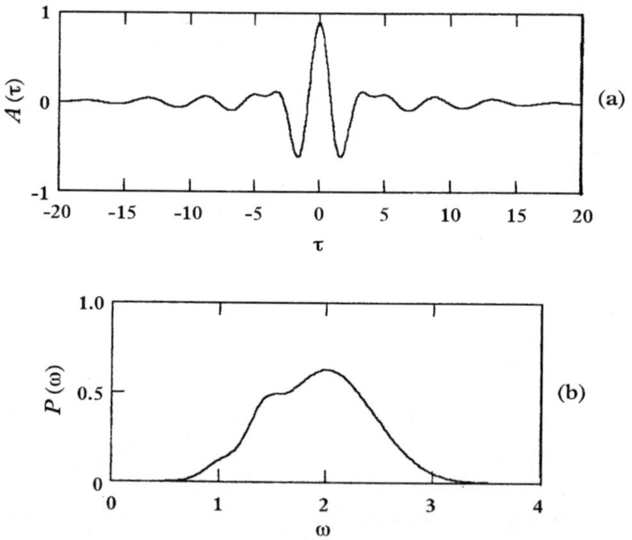

Figure 1 (a) An example of an auto-correlation function for a collimated thermal signal beam and (b) the Fourier transform of this auto-correlation function which is the spectrum of the beam. The form of the spectrum has been arbitrarily chosen for the purposes of illustration. In particular, the units for the angular frequency ω, and those for the time delay τ, are inversely related but are otherwise arbitrary; for example they might be s^{-1} and s, respectively, or $10^{12}s^{-1}$ and ps, etc.

The power spectral density function, *P*(*ω*)

The statistical properties of a beam can alternatively be expressed *in the frequency domain* by the Fourier transform of the autocorrelation function, i.e. by

$$P(\omega) \equiv \int_{-\infty}^{+\infty} A(\tau) \exp - i\omega\tau \, \mathrm{d}\tau$$

Since $A(\tau)$ is necessarily symmetrical about $\tau = 0$, this transform can be written as the cosine transform

$$P(\omega) \equiv \int_{-\infty}^{+\infty} A(\tau) \cos\omega\tau \, \mathrm{d}\tau$$

$P(\omega)$ is known as the *power spectral density function* of the beam (this is a signal power density, in the sense indicated above, i.e. the true power density in watts per hertz is given by multiplying the signal power density by the cross-sectional area of the beam and by the impedance of free space). The physical significance of the power spectral density, $P(\omega)$, can be understood as follows. A collimated thermal signal beam can be treated as a superposition of coaxial collimated beams the frequencies of which span a continuous spectrum, and the amplitudes and phases of which fluctuate randomly in time but with definite mean values. The power spectral density, $P(\omega)$, expresses the distribution of signal power over angular frequency, ω, i.e. $P(\omega)\Delta\omega$ is the signal power carried by all those waves in the superposition which have frequencies between ω and $(\omega + \Delta\omega)$, where $\Delta\omega$ is a small increment in ω (see Ref. 5 for an explicit demonstration of this Wiener–Khinchin theorem).

The term *PSD spectrum*, or simply *spectrum*, will, as is customary, be used as a shorter form for power spectral density function.

Thus the autocorrelation function (ACF) and the power spectral density function (PSD) of a signal beam are alternative, not independent, ways to specify a thermal beam's primary statistical properties. The PSD of any signal beam could be obtained from its ACF by computing the Fourier transform above, or the ACF obtained from the PSD using the inverse Fourier transform. The spectrum of the beam whose autocorrelation function is shown, as an example, in Figure 1(a) is given in Figure 1(b), as obtained by Fourier transformation.

The Fourier transform relationship between the two indicates, *inter alia*, that the value of $A(\tau)$ at $\tau = 0$ is $\langle E^2(t) \rangle$, which is the total signal power in the beam and therefore equal to the integration of $P(\omega)$ over all ω. Correpondingly, the integral of $A(\tau)$ over all τ is equal to the power in the spectrum at $\omega = 0$ and in practice that will be zero. The maximum rates of change of the E-field at a point in the beam, i.e. the maximum slopes in $E(t)$, relate to the upper bound of the PSD spectrum.

Change in power spectral density on linear filtering

If a collimated linearly polarized thermal signal beam is passed through a linear optical system which maintains its collimation and linear polarization, its power spectral density function or spectrum will generally be modified; it will be frequency filtered in some way. The character of the modification can be expressed in terms of the frequency response function, or transfer function, of the linear system, $H(\omega)$ say, which is defined by reference to the passage of coherent beams through the system. A collimated coherent signal beam of angular frequency ω which enters a linear time-invariant system emerges as a coherent beam of unchanged frequency ω, but with reduced amplitude ($A_i \rightarrow A_e$, say) and shifted phase ($\omega t \rightarrow \omega t + \theta$, say). The ratio A_e/A_i, and the phase shift θ, will depend on ω but are independent of the signal amplitude. A_e/A_i is the modulus and θ the phase-angle of the system's complex 'frequency response function' or 'transfer function' $H(\omega)$.

If a collimated linearly polarized thermal signal beam is passed through a linear optical system, its PSD spectrum is modified according to

$$P_e(\omega) = |H(\omega)|^2 P_i(\omega)$$

where $P_i(\omega)$ is the PSD spectrum of the incident beam, $P_e(\omega)$ that of the emergent beam and $H(\omega)$ is the transfer function of the linear system. This is a central result of linear systems analysis (a derivation is given in Refs 5 and 6).

The cross-correlation function and cross-spectral-density function of two thermal signal beams

The auto-correlation function of a thermal beam measures the correlation between the E-field in the beam at one instant in time with the field in the same beam a time interval τ later. The cross-correlation function (CCF) of two superposed collimated thermal beams measures the correlation between the two beams, also as a function of time interval τ. It is

$$A_{12}(\tau) \equiv \langle E_1(t+\tau)E_2(t)\rangle$$

where E_1 and E_2 are the E-fields of the two beams at a given point.

The cross-spectral-density function (CSD) of the two beams is the Fourier transform of their cross-correlation function:

$$P_{12}(\omega) \equiv \int_{-\infty}^{+\infty} A_{12}(\tau) \exp - i\omega\tau \, d\tau$$

The change in the cross spectral density on linear filtering

If a thermal beam is divided into two beams which are then passed through different linear filters and superposed again, the cross spectral density of the superposed final beams, $P_{12}(\omega)$, is related to the power spectral density of the initial beam, $P_i(\omega)$, thus:

$$P_{12}(\omega) = H_1(\omega)H^*{}_2(\omega)P_i(\omega)$$

where $H_1(\omega)$ and $H_2(\omega)$ are the complex transfer functions of the two filters [5, 6].

2 THE TWO-BEAM INTERFEROMETER AS SPECTRUM ANALYSER

2.1 POLARIZATION-DIVISION (MARTIN–PUPLETT) INTERFEROMETERS

A two-beam interferometer is made up of two optical four-ports in tandem, as shown in Figure 2. The first of these is known as the beam divider, and the collimated signal beam to be analysed by the interferometer enters this through one of its ports (D1, say). The beam is divided there into two collimated beams which emerge from ports D3 and D4. The two beams travel different paths, via reflectors R1 and R2, repectively, to the second four-port, known as the beam combiner, one entering at port C1 and the other at port C2. Each of these incident beams is divided in the combiner into two further beams, one of which emerges from port C3 and the other from port C4. The collimated beams that emerge from ports C3 and C4 are therefore both superpositions of two constituent beams which had travelled different paths in the interferometer, one via reflector R1 and the other via R2. A second signal beam could be directed into port D2 of the beam divider and would similarly deliver superposed output beams at C3 and C4. The beams emerging from C3 and C4 would then both be superpositions of four constituent beams. An interferometer is thus itself a four-port, having two input ports (D1 and D2) and two output ports (C3 and C4).

The beam paths via R1 and R2 inside the interferometer differ in length and the difference can be changed by displacing R2 along the common direction of its incoming and returning beams. The relative phases of the superposed beams in each of the two output ports change with changing path difference and the mutual interference between them results in a changing total power carried by the beam.

A detector placed in either output beam will record the variation of signal power with changing path-difference as an ‘interferogram.’ Our purpose in the following sections will be to show that an appropriate interpretation of such an interferogram provides a spectral analysis of the incident signal beam. The matters of immediate interest are the structure and function of the beam divider and beam combiner.

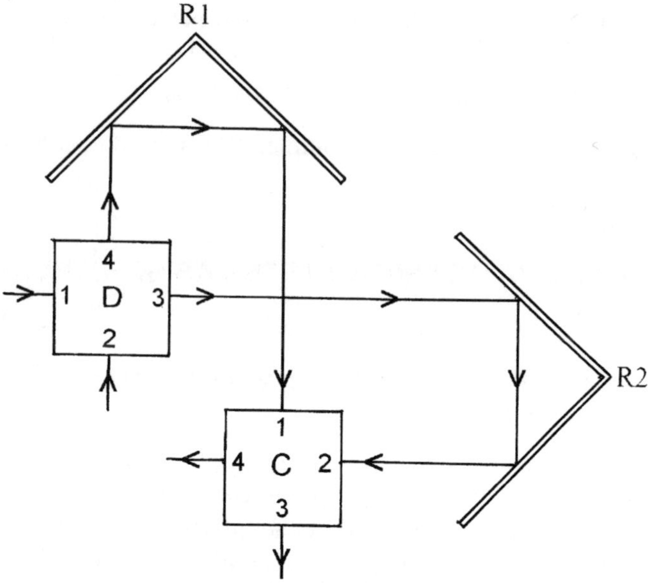

Figure 2 A two-beam interferometer illustrated as two optical four-ports in tandem; the first four-port (D) is known as the beam divider and the second (C) as the beam combiner. A collimated signal beam which is incident at input port D1 is divided in D into two beams, one of which leaves by port D4 and propagates via the pair of reflectors R1 to the port C1 of the combiner C, whilst the other leaves by port D3 and propagates to port C2 of C via the reflector pair R2. Each of the two beams incident on C is further divided in C, to contribute a beam leaving C at port C3 and another leaving by port C4. The beam incident at port D1 therefore results in beams emerging from output ports C3 and C4, each of which is a superposition of one beam which travelled through the interferometer via R1 and another which travelled via R2. The same must be true also for a collimated beam incident on the divider at port D2. If collimated signal beams are incident on both D1 and D2, the beam emerging from port C3 and that emerging from C4 will each be a superposition of four beams. The power carried by each emergent beam will depend on the interference between its contributory beams. Movement of one of the reflector pairs in the interferometer (say R2) along the line of incidence will change the difference in the path length of the beams travelling via R1 and R2; the interference between the contributory beams changes as a consequence and therefore also the power carried by each emergent beam. The variation in power recorded by a detector system placed in either emergent beam, as the path difference is progressively changed, is known as an interferogram

The four-ports used in Michelson interferometers give amplitude division and combination by the ordinary processes of reflection and transmission at a flat sheet of dielectric material. The four-ports in a Martin–Puplett interferometer give amplitude division and combination by resolution of orthogonal linear polarizations at flat wire grids; this is commonly called polarization division (and combination) and is examined in this section.

The internal structure of a polarization-division divider is the same as that of a combiner. Each contains two flat grids of parallel conducting wires (Figure 3). A collimated coherent signal beam incident on a planar wire grid would be reflected and transmitted into many orders of diffraction if the spacing between the wires, s, were greater than the wavelength $\lambda \equiv c/\nu$ (where ν is the frequency of the signal beam and c the velocity of light). If, however, $s < \lambda$, only zero-order diffracted beams can be generated, i.e. an undeviated transmitted beam and a beam specularly reflected from the plane of the grid. If, moreover, $s \ll \lambda$, these reflected and transmitted beams will be essentially purely plane-polarized, regardless of the state of polarization of the incident beam. It is this function of a grid which is required in a beam divider/combiner.

To see why a grid with $s/\lambda \ll 1$ should function in this way, consider first a collimated linearly polarized beam incident in a direction perpendicular to the wires and with its E-field parallel to the wires [Figure 4(a)]. The beam will induce electrical currents in the wires and, when $s \ll \lambda$, the pattern of current flow in the plane of the grid will be essentially the same as would be induced if a plane metallic sheet replaced the grid; this implies that, for this polarization of the incident beam, the grid will reflect most of the incident beam power specularly from its plane, and will transmit little. If, instead, the incident beam

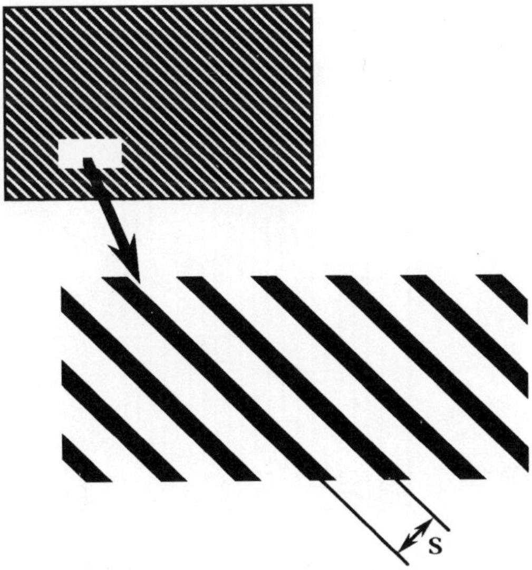

Figure 3 To illustrate one of the flat grids of parallel conducting (metallic) wires used in a beam divider/combiner; *s* is the spacing of the wires

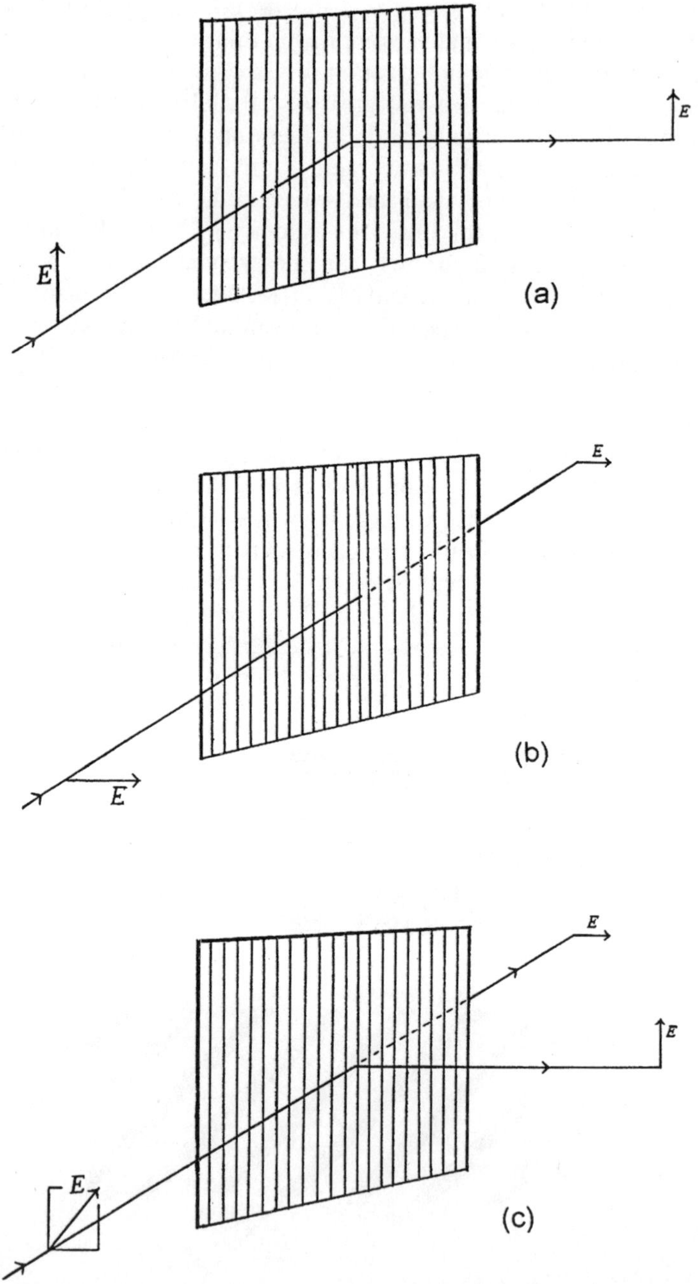

Figure 4 To illustrate polarization-division by a flat wire grid as described in the text

were linearly polarized with its E-field perpendicular to the grid-wires [Fig. 4(b)], an electrical polarization would be induced across the width of the wires, with no current flow along the wires (the word polarization here refers to dielectric polarization of a material, not to polarization of an electromagnetic wave). The grid would thus behave in this case as would a thin dielectric sheet; that implies high transmission for this polarization, and low reflection [Figure 4(b)].

If the incident coherent beam had an arbitrary state of polarization, the grid would resolve it into a specularly reflected beam polarized parallel to the wires and a transmitted beam polarized perpendicularly to the wires [Figure 4(c)].

Similar conclusions in fact follow for a beam incident along a line having an arbitrary orientation relative to the wires [Figure 5(a)]. If the incident beam is linearly polarized with its E-field in the plane containing both the direction of incidence and the line of the wires, the field will have a non-zero component along the wires so current will flow in the wires and the result is a reflected beam linearly polarised with its E-field in the plane containing both the direction of reflection and the line of the wires. An incident beam having the orthogonal polarisation has its E-field perpendicular to both the direction of incidence and the line of the wires in the grid, so it induces no current along the wires and the result would be a transmitted beam linearly polarized with its E-field perpendicular to the plane containing both the line of transmission and the line of the wires [Figure 5(b)].

The general inference is, therefore, that an incident coherent beam of arbitrary polarization state will be resolved by a grid, with $s/\lambda \ll 1$, into a transmitted beam and a specularly reflected beam both of which are linearly polarized, the former with its E-field perpendicular to the plane containing both the direction of transmission and the line of the wires, and the latter with its E-field in the plane containing both the direction of reflection and the line of the wires [Figure 5(c)].

The conclusion drawn above applies when $s/\lambda \ll 1$. The patterns of current and of polarization induced in the grid are spatially modulated with period s and the interference between the field of the incident beam and the fields radiated in the transmit and reflect directions by the induced current and polarization will only be complete if $s/\lambda \rightarrow 0$. If s/λ approaches 1, that interference will not be complete and the transmit and reflect beams will not be perfectly plane-polarized; each will include a secondary cross-polarized component. The closer s is to λ, the greater will be the leaked cross-polar components.

As noted above, if $s/\lambda > 1$, the incident signal power would be diffracted into multiple diffraction orders in transmission and reflection and the grid could not serve as a beam divider.

Figure 6 shows an example of the results of a theoretical calculation of the transmittance of a grid for normally incident beams of each linear polarization,

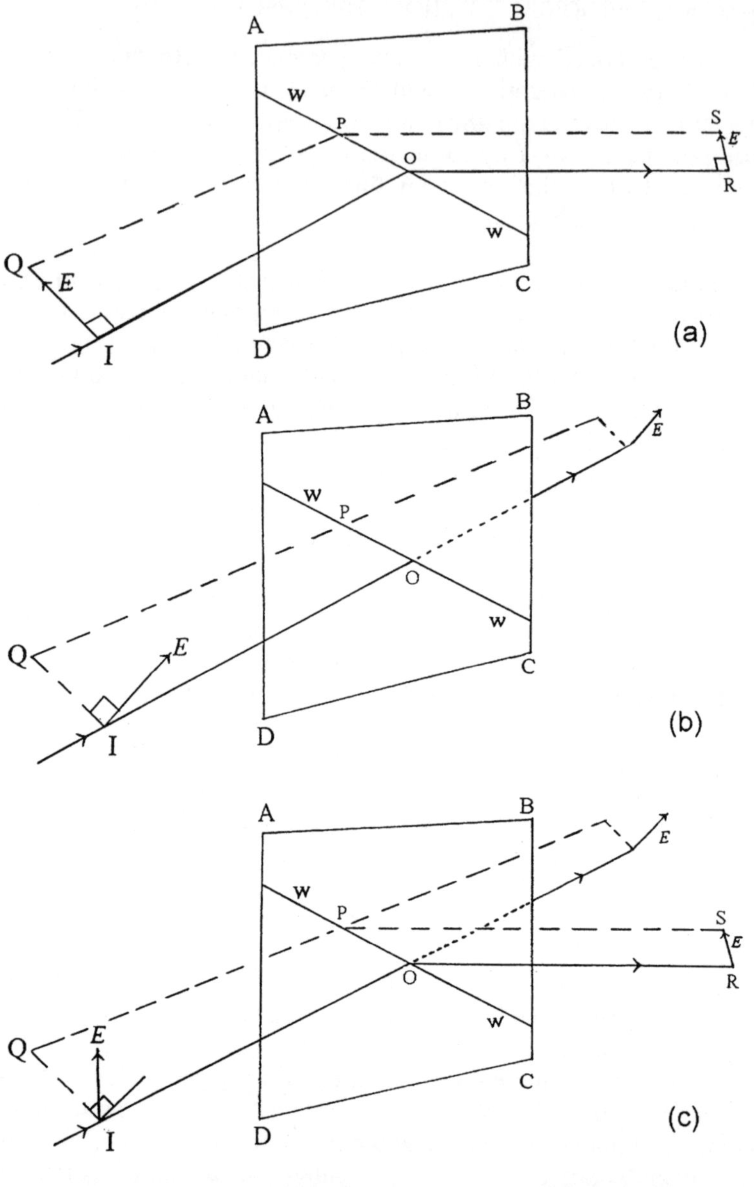

Figure 5 To illustrate (a) the specular reflection by a wire grid of an incident beam linearly polarized in a direction having a component of the electric field, *E*, along the line of the wires ww, (b) the transmission through a wire grid of an incident beam having a zero component of *E* along the line of the wires, ww, and (c) the resolution of an incident linearly polarized beam into orthogonally polarized components by a wire grid. ABCD is the plane of the grid, IOPQ is the plane containing both the direction of incidence of the beam and the line of the wires and ORSP is the plane containing both the direction of specular reflection and the line of the wires

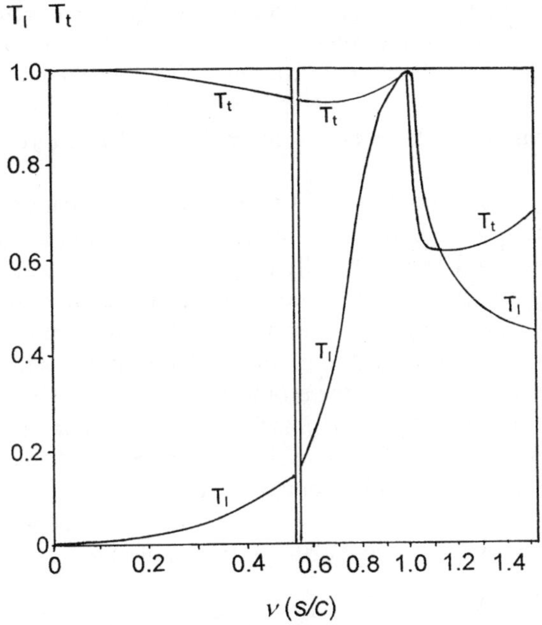

Figure 6 The theoretically derived variation with signal-beam frequency of the power transmittance of a flat wire grid for a coherent, normally incident, signal beam; T_t is the transmittance for a beam linearly polarized with its *E*-field perpendicular to the line of the wires in the grid, and T_l is that for a beam polarized parallel to the wires; ν is the signal frequency, *s* is the spacing of the wires in the grid and *c* is the velocity of light. This figure is based on the results reported in Ref. 7. The wires were assumed in the analysis to be of circular cross-section, and the wire-spacing to be 3.5 times the wire diameter

as dependent on s/λ (and for the shapes and the relative dimensions of the wires and gaps referred to in the caption). This confirms that efficient polarization division is obtained with $s/\lambda \ll 1$, with a progressive fall in efficiency as s/λ approaches 1. Other theoretical analyses for grids having various shapes of wire and gap, and for various directions of incidence, lead to curves of the same general form as that in Figure 6 but with some variation in detail in the range above $s/\lambda \approx 0.3$. A real grid will not have a perfectly uniform spacing of the wires and non-uniformities will have some influence on the polarizing properties in the same range of s/λ.

For the purposes of describing the structures of Martin–Puplett or polarization-division interferometers in this section, it will be convenient first to describe what happens to *coherent* beams having a frequency for which the grids have full polarization-division efficiency, as they pass through the system.

Polarization-division beam dividers and beam combiners are identical in structure. In each there are two grids arranged as shown in Figure 7. The beams entering, passing through and leaving the divider/combiner lie in a common plane of incidence. The directions of the wires in the two grids are orthogonal, i.e. their projections on to a plane perpendicular to the beams' propagation direction between the polarizers are at $\pi/4$ and $3\pi/4$ rads to the plane of incidence respectively.

One input port of a beam divider/combiner admits a beam plane-polarized in the direction wholly transmitted by the first grid, and the other admits a beam plane-polarized in the direction wholly reflected from the grid. If a signal beam arriving at an input port is not plane-polarized in the required direction, the grid will resolve it into an admitted beam and an orthogonally polarized component which is diverted into a power dump (Figure 8); only the admitted component will then be analyzed by the interferometer, of course. (It will be shown in Section 2.4 how the divider in a polarization-division interferometer can be modified to allow the analysis of an arbitrarily polarized incident beam.)

The two admitted beams are incident in directions such that they emerge from the first grid superposed with a common direction of propagation. At this point the superposed beams are orthogonally polarized. They encounter the

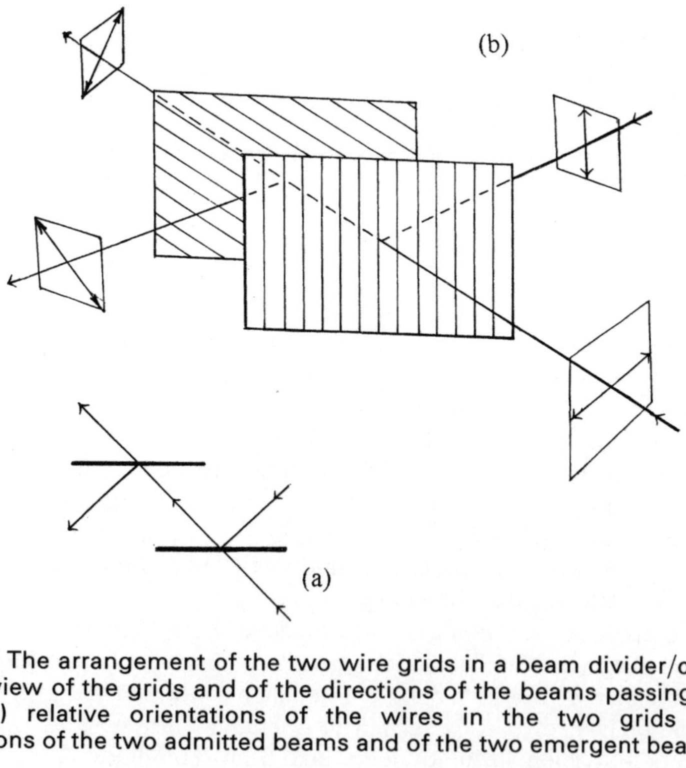

Figure 7 The arrangement of the two wire grids in a beam divider/combiner. (a) Plan view of the grids and of the directions of the beams passing through them; (b) relative orientations of the wires in the two grids and the polarizations of the two admitted beams and of the two emergent beams

second grid, which serves to split both equally between the two output ports of the divider/combiner and to superpose them with the same plane of polarization but as a vector sum in one port and a vector difference in the other. The diagram in Figure 9 shows why this is. The instantaneous field amplitudes of the orthogonally polarized beams incident on the second grid combine vectorially to produce a component along the projected line of the wires which is $1/\sqrt{2}$ times the *sum* of the incident amplitudes, and a component transverse to the wires which is $1/\sqrt{2}$ times the *difference*. The induced currents and polarizations in the wires therefore produce a beam reflected from the grid with amplitude equal to $1/\sqrt{2}$ times the sum of the amplitudes of the admitted beams, and a beam transmitted through the grid with amplitude equal to $1/\sqrt{2}$ times the difference.

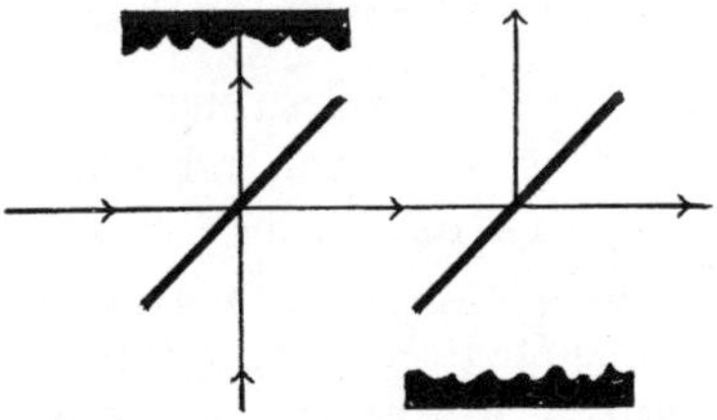

Figure 8 The component of an incident beam which is polarized orthogonally to that admitted by a beam divider/combiner will be deposited in a power dump, as illustrated schematically here

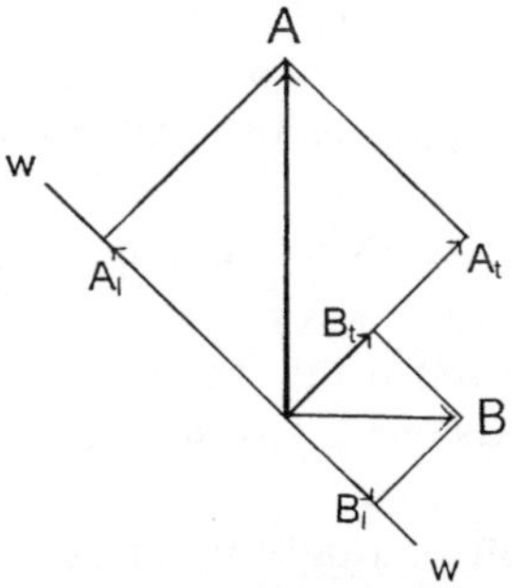

Figure 9 To illustrate the combination of the polarization vectors of two orthogonally polarized beams incident on a wire grid. A and B denote the instantaneous *E*-fields of the two beams. The currents and (dielectric) polarizations induced in the grid will be determined by the components of the vector resultant of A and B along, and transverse to, the line of the wires in the grid (ww), respectively. The components of A and B along the direction of the wire, A_l and B_l, can be seen here to be *oppositely* directed. The components of A and B transverse to the direction of the wires, A_t and B_t, are, however, *similarly* directed

The divider and combiner are configured in an interferometer so that the beams emerging from ports D3 and D4 of the divider have the polarizations required for admission at the input ports of the combiner, C2 and C1 respectively.

With the relationships illustrated in Figure 9 in mind, it is possible to write directly expressions for the complex amplitudes of the signal beams leaving the two output ports of an interferometer when two coherent beams, of angular frequency ω, are incident in the two input ports. If the complex amplitudes of the beams entering ports D1 and D2 are E_1 and E_2, respectively, the complex amplitudes of the beams leaving the ports C3 and C4 are as follows:

$$\text{port C3:}\quad \tfrac{1}{2}(E_1 + E_2)\exp i\omega(t+\tau) + \tfrac{1}{2}(E_1 - E_2)\exp i\omega t$$

$$\text{port C4:}\quad \tfrac{1}{2}(E_1 + E_2)\exp i\omega(t+\tau) - \tfrac{1}{2}(E_1 - E_2)\exp i\omega t$$

where τ is the difference between t_L and t_s, the times taken to travel from the input ports to the output ports by the longer and shorter paths, respectively. (To simplify the expressions, the origin of t for the fields in the output ports has been shifted by t_L from that for the input ports; time averages will be the quantities of interest because detectors integrate over time, so this is of no consequence.)

The signal beams involved in spectral analysis are wide-band thermal signals, not coherent beams. To deal with such beams it is necessary to go to the time domain in first instance rather than the frequency domain. The E-field at a point in such a signal beam fluctuates randomly in time. There will be an upper limit to the rate of change of the field, however, and that can be viewed as an upper limit to the spectral frequencies included in the beam. The resolution of the field at a grid into orthogonal polarization components, as illustrated in Figure 9, will apply to the instantaneous directions of the field amplitudes in the incident and emergent beams at the grid *provided* that the grid has full polarization efficiency at all frequencies up to the upper limit, i.e. provided that the grids are 'ideal.' In continuing the description of Martin–Puplett interferometers in this section, and in the next section, ideal grids in this sense will be assumed. Consideration is given to interferometers with non-ideal grids in Section 2.3.

With ideal grids, then, fluctuating noise fields $E_1(t)$ and $E_2(t)$ incident at the input ports D1 and D2 will lead to the following fluctuating fields in the output ports C3 and C4;

$$\text{port C3:}\quad \tfrac{1}{2}[E_1(t+\tau) + E_2(t+\tau)] + \tfrac{1}{2}[E_1(t) - E_2(t)]$$

$$\text{port C4:}\quad \tfrac{1}{2}[E_1(t+\tau) + E_2(t+\tau)] - \tfrac{1}{2}[E_1(t) - E_2(t)]$$

This is a time-domain statement of the overall function of an interferometer with ideal grids. It is the starting point for the treatment of spectrum analysis in the next section.

Before turning to that, however, a way to simplify significantly the structure of a polarization-division interferometer by merging the divider and combiner into one unit should be described.

Consider first the much simplified structure illustrated in Figure 10. The two beams inside the interferometer are returned to the divider by plane reflectors. The divider then serves also as combiner. The input port D1 and the output port C4 coincide, however, and so too do the input port D2 and the output port C3. That could give rise to problems in applications involving weak signal powers and cooled detectors. A detector placed in the beam emerging from C3 will record an interferogram but it becomes the source for a thermal signal entering the input port D1 which will contribute to (contaminate) the interferogram—the detector sees itself. (The simplest, and most widely used, type of Michelson interferometer similarly denies access to two of its ports.)

Access to all four ports can be retained, however, in a compact configuration, if a wide-band $\pi/2$ polarization rotator can be introduced into each of the two internal beams. Most polarization rotators are narrow banded but the extremely simple type of rotator introduced in the first Martin–Puplett interferometers, and used in many such interferometers since, operates well from microwave to near-infrared frequencies. It is a $\pi/2$-roof-edge metallic retro-reflector set with its roof-line at $\pi/4$ to the plane of polarization of the incoming beam; the plane of polarization of the returning beam is rotated through $\pi/2$ from that of the incoming beam for the reasons given in Figure 11. With such a rotator in each internal beam a single grid can serve both as the second grid of the divider and as the first grid of the combiner and, as shown in Figure 12, the four ports are again separated. The earlier statement of the function of a polarization-division two-beam interferometer applies to this type of interferometer.

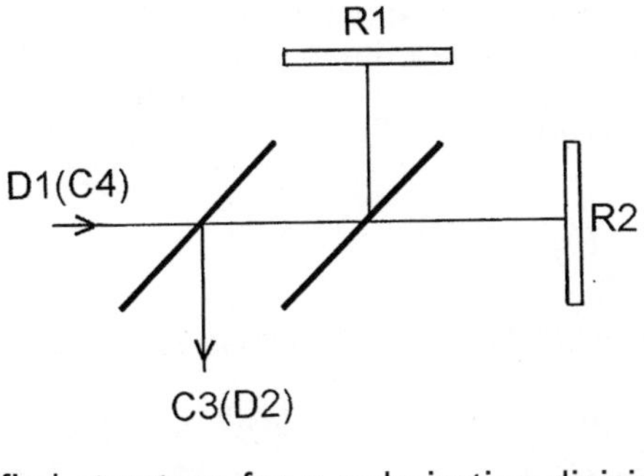

Figure 10 A simplified structure for a polarization-division interferometer in which the functions of beam divider and beam combiner are combined in one two-grid unit. The input port D1 and the output port C4 are coincident in this structure, as are the output port C3 and the input port D2

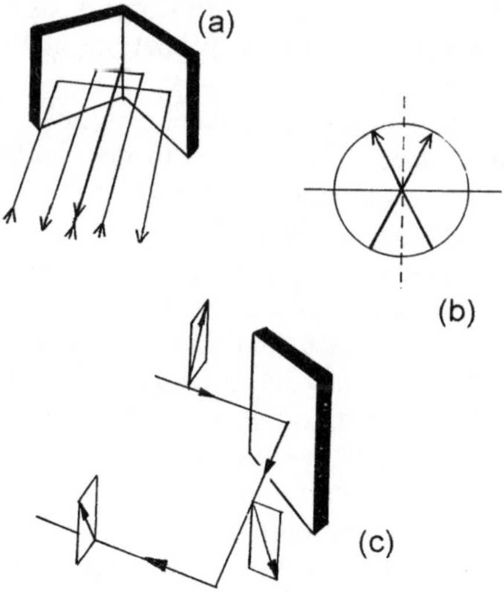

Figure 11 A linearly-polarized signal beam, incident on a roof-top reflector with a $\pi/2$ roof-angle, in a direction perpendicular to the roof-edge, is returned along the same line [as is indicated by the three illustrative rays shown in (a)]. The plane of polarization of the returned beam differs, however, from that of the incident beam; it is rotated to a direction making an equal angle to the normal to the plane of incidence but on the opposite side, as illustrated in (b). This is a consequence of the electromagnetic boundary condition that the electric field at a point on the surface of a metal must be normal to the surface. To satisfy this boundary condition at the first face of the reflector, the polarization direction of the beam reflected from the first surface must be related to that of the incident beam as indicated in (c). Similarly, to satisfy the boundary condition at the second face, the relative polarizations of the incident and emergent beams there must be as indicated in (c) [in the interest of diagrammatic clarity, the second face itself is not drawn in (c), only the incident and reflected beams]. The overall effect of the two reflections at the roof-edge reflector is therefore as described above. For the special case in which the polarization direction of the incident beam is at $\pi/4$ to the roof-edge, the rotation of the plane of polarization is through $\pi/2$; this is the case for the roof-top reflectors used in Martin–Puplett interferometers (Figure 12)

2.2 SPECTRUM ANALYSIS (WITH IDEAL GRIDS)

A two-beam interferometer with ideal grids, as described in the preceding section, serves as a spectrum analyser because, as will be seen below, an interferogram produced by such an interferometer is the auto-correlation function (ACF) of the incident signal beam and the power-spectral-density spectrum of the signal beam can therefore (Section 1.2) be obtained by

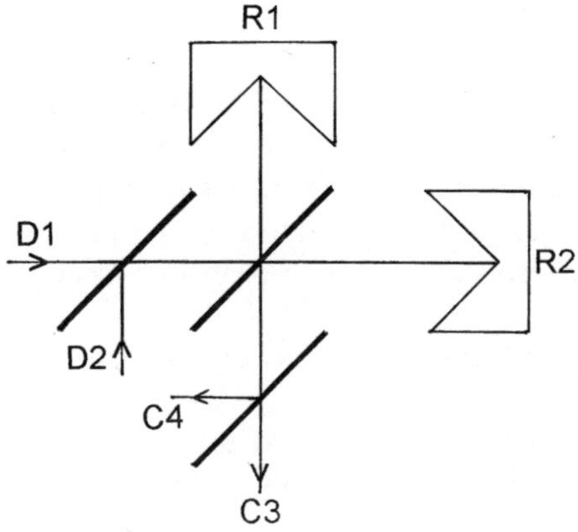

Figure 12 To illustrate a Martin–Puplett interferometer incorporating roof-top retroreflectors, R1 and R2, which serve to rotate the plane of polarization of a signal beam through $\pi/2$, and three wire grids serving as beam divider/combiner. The input ports of the interferometer, D1 and D2, and the output ports, C3 and C4, are separated and accessible

computing the Fourier transform of the interferogram/auto-correlation function.

An interferometer goes straight to the definition of the auto-correlation function of a signal beam (Section 1.2). Each of the two output beams from an interferometer having ideal grids is a superposition of two beams superposed with a relative time shift, $\tau = d/c$, where d is the difference in the lengths of the paths followed by the separated beams inside the interferometer and c is the velocity of light. The total field at a point in the composite beam is (Section 2.1)

$$\tfrac{1}{2}[E_1(t+\tau) + E_2(t+\tau)] \pm \tfrac{1}{2}[E_1(t) - E_2(t)]$$

where $E_1(t)$ and $E_2(t)$ are the time-dependent fields in the collimated linearly polarized beams admitted at input ports D1 and D2 of the interferometer. The $\pm$ sign alternatives apply to the fields in the two output ports, C3 and C4, respectively.

Now, the output from a detector incorporating an integrator is proportional to the signal power carried by the beam incident on the detector, averaged over the integration time of the detection system (it will be assumed in this section that the detector's responsivity is independent of signal frequency—this assumption will be corrected in the next section). The output from the detector in the interferometer is therefore proportional to the square of the expression above averaged over time, i.e.

$$\tfrac{1}{2}\{\langle[E_1(t)]^2\rangle+\langle[E_2(t)]^2\rangle\}\pm\tfrac{1}{2}[\langle E_1(t)E_1(t+\tau)\rangle-\langle E_2(t)E_2(t+\tau)\rangle]$$

where the angle brackets $\langle\ \rangle$ indicate averages over the detection integration time and cross-terms $\langle E_1(t)E_2(t)\rangle$, $\langle E_2(t+\tau)E_1(t+\tau)\rangle$, $\langle E_1(t)E_2(t+\tau)\rangle$ and $\langle E_2(t)E_1(t+\tau)\rangle$ are omitted because the two input beams, originating in different thermal sources, are uncorrelated (correlated inputs are considered in Section 2.4). An interferogram is a record of the variation of this quantity with the time delay τ.

The first term is the sum of the powers of the two input signals and is independent of τ; it provides a DC level. The second term is the difference of the auto-correlation functions of the two input signal beams. (Strictly, auto-correlation functions involve averaging over infinite time; however, the detector integration times involved in practice—usually several microseconds or more—are long compared with the correlation times of thermal signal beams and suffice to give time averages indistinguishable from those that would be obtained with longer integration [5].)

The interferogram from an ideal interferometer is thus the difference of the auto-correlation functions of the two input signal beams, superposed on a DC level equal to the sum of the powers in the two beams.

The ± sign alternative above refers to the two output beams, so the interferograms in the two output ports are complementary (Figure 13). The sum of the two output powers is constant independent of τ and equal to the sum of the two input powers, as might be expected. As the moveable reflector in the interferometer is displaced, the total input power is divided between the two output ports.

Computation of the Fourier transform (FT) of the ACF, i.e. of either of the output interferograms, will give the difference in the PSD spectra of the two input beams (Section 1.2).

FT of interferogram in outport port C3 gives: $\tfrac{1}{2}[P_2(\omega)-P_1(\omega)]$

FT of interferogram in outport port C4 gives: $\tfrac{1}{2}[-P_2(\omega)+P_1(\omega)]$

where $P_1(\omega)$ and $P_2(\omega)$ are the PSD spectra for the input beams in ports D1 and D2, respectively.

An interferometer is thus both an auto-correlator and a spectrum analyser. It is, however, a *differential* spectrum analyser, measuring the difference of the spectra of the two input signals. There will always be some signal power in the second port; if none is explicitly delivered to the port, there will be the thermal signal from objects at ambient temperature in the path of the input. If no signal beam is explicitly delivered to either input port so that both are filled with thermal radiation from objects at the same ambient temperature, a null spectrum will be recorded. The presence of ambient temperature input

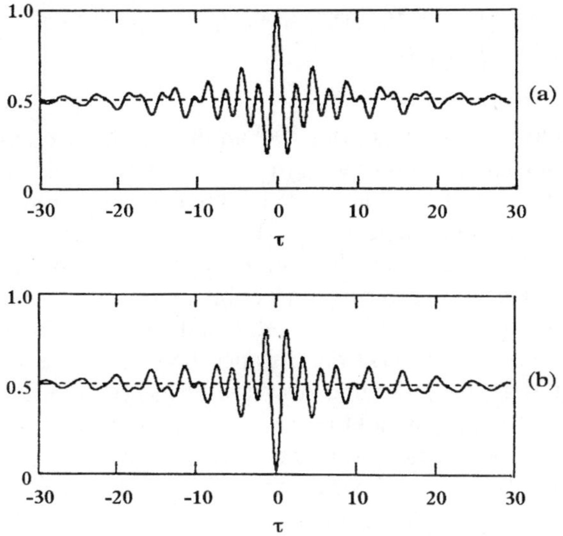

Figure 13 The interferograms produced in the two output ports of an ideal interferometer will exhibit the complementary character of the examples shown here in (a) and (b), i.e. the sum of the two is constant over all time delays, τ. (The detailed form of these illustrative examples and the units of the time delay, τ, are arbitrary)

to the second port often goes unremarked in applications of interferometric spectrometers (in fact the very existence of a second input port is frequently unappreciated); that will lead to error unless the input signal delivered to the first port is much larger than the ambient temperature input in the second.

The detector will radiate thermally into the interferometer and this signal power will emerge from the input ports. If some of this signal power were to be reflected back, to return through the interferometer to the detector, it would contaminate the interferogram (and, having suffered the path difference twice, its contribution would be at twice its true frequency). Such reflection must be avoided especially when a cryogenic detector is in use; if it is, the temperature of the detector will be of no consequence.

In practice, of course, the path difference in an interferometer will have an upper limit so the auto-correlation function will be determined up to the corresponding maximum value of τ only. Also, an interferogram will be sampled at discrete intervals of path difference, not continuously. The Fourier transform will therefore be computed as a truncated and discrete transform; the implications of this for measured spectra are examined in Sections 2.4 and 2.7.

2.3 *SPECTRUM ANALYSIS (WITH NON-IDEAL GRIDS AND NON-IDEAL DETECTOR)*

Ideal wire grids were assumed in the preceding section. The performance of an interferometer having non-ideal grids is examined in this section. It will be seen that an interferometer with non-ideal grids can function usefully as a spectrum analyser over a range of frequency, provided that a calibration is effected.

As previously, the time-dependent E-field at a point in a linearly polarized collimated incident signal beam is denoted $E(t)$. If the beam follows a path through a train of grids, suffering reflection or transmission at each, the E-field at a point in the emergent collimated beam would, if the grids were ideal, be $\frac{1}{2}E(t+T)$, where $T = D/c$ (D the distance between the two points and c is the speed of light. That is, the time structure of the field in the output beam is the same as that of the input field apart from the propagation time delay, T. If the grids were non-ideal, however, the time structure of the beam would be modified, not simply delayed, because the more rapid variations in time will be damped owing to the imperfect polarization properties of the grids at high frequencies. There is an implied loss of signal power from the beam and this lost power appears as a leak from the point indicated in Figure 14. The emergent time-dependent signal field will be denoted $E'(t)$ and its relationship to $E(t)$ be examined later.

The symmetry of the interferometer is such that the two paths, variable and fixed, that link a given pair of ports involve the same number of transmissions and of reflections at the grids; the implication of this is that the time structures of the collimated beams that follow the two paths will be the same apart from the time delay associated with the path difference, i.e. they can be written $E'(t)$ and $E'(t+\tau)$, respectively, where τ is time delay arising from the path

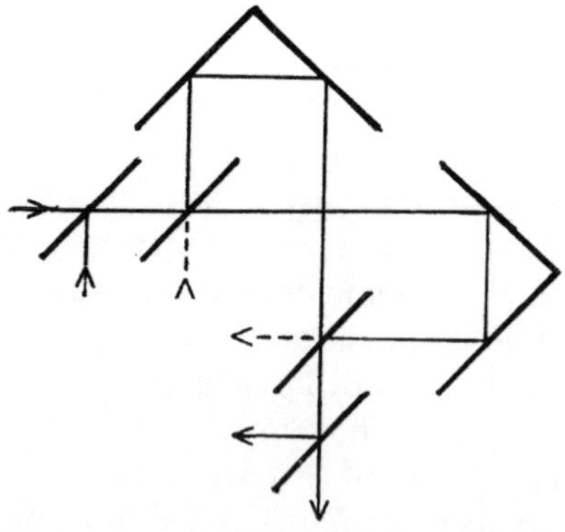

Figure 14 Some signal power is lost from an interferometer, at the points indicated here by broken lines, when the grids in the interferometer are imperfect polarization dividers

difference, d, i.e. $\tau = d/c$. The line of argument used in Section 2.2 now leads to the following expressions for the variation of the signal power in an output port with the path difference (omitting terms independent of the path difference):

In output port C3: $\langle E'_{13}(t)E'_{13}(t+\tau)\rangle - \langle E'_{23}(t)E'_{23}(t+\tau)\rangle$

In output port C4: $-\langle E'_{14}(t)E'_{14}(t+\tau)\rangle + \langle E'_{24}(t)E'_{24}(t+\tau)\rangle$

The suffixes here indicate the input and output port involved in the pertinent beam-path (e.g. $E'_{13}(t)$ and $E'_{13}(t+\tau)$ are the time-dependent E-fields for beams travelling from port D1 to port C3 via the fixed path and via the variable path, respectively.

These functions of τ are the interferograms in the two ports. It can be seen that the first is the difference of the auto-correlation functions of $E'_{13}(t)$ and of $E'_{23}(t)$, and the second the difference of the auto-correlation functions of $E'_{14}(t)$ and of $E'_{24}(t)$. The Fourier transforms of the interferograms are therefore the differences of the power spectral density functions (the spectra) of the output beams.

The interferometer thus measures directly the PSD spectra of the output fields, not those of the input fields. For an interferometer having ideal grids, the distinction between output and input beams does not arise because the time structure of the E-field of a beam does not change on propagation through ideal grids. For an interferometer with non-ideal grids there is a distinction, however, and since the purpose of spectrum analysis is a determination of the spectra of incident beams, it is necessary to know how the spectrum of a signal beam changes on propagation through a train of non-ideal grids.

A train of non-ideal grids is a linear time-invariant system and, as noted in Section 1.2, the spectrum therefore changes in accordance with

$$P'(\omega) = |H(\omega)|^2 P(\omega)$$

where $P(\omega)$ is the PSD spectrum at the input to the system, $P'(\omega)$ is that at the output and the power-transmission function, $|H(\omega)|^2$, is the squared modulus of the system's frequency-response function, or transfer function, $H(\omega)$.

The Fourier transforms (FT) of the interferograms of the two output ports thus provide the following measures:

The FT of the interferogram in output port C3 gives:

$$|H_{13}(\omega)|^2 P_1(\omega) - |H_{23}(\omega)|^2 P_2(\omega)$$

The FT of the interferogram in output port C4 gives:

$$-|H_{14}(\omega)|^2 P_1(\omega) + |H_{24}(\omega)|^2 P_2(\omega)$$

where $P_1(\omega)$ and $P_2(\omega)$ are the PSD spectra of the signal beams incident in input ports D1 and D2, respectively, and the several power-transmission functions refer to the paths through the interferometer linking an input port (1 or 2) with an output port (3 or 4) as indicated by the suffixes. (There is a small contribution to the output additional to those above arising from the fact that non-ideal grids will inject some signal power due to thermal emission from the grids themselves or from other parts of the interferometer in their vicinity. Consideration is given to this small additional contribution when examining radiometric measurements in Section 3.1.)

The power-transmission function of a train of grids, $|H(\omega)|^2$, will be near the ideal value of $\frac{1}{2}$ for all frequencies below about $2c/s$ where s is the wire-spacing (Section 2.1). It will fall in magnitude progressively as the frequency rises beyond that value, and will reach low values for frequencies near and above $2\pi c/s$. The power lost is primarily due to leakage from the system, notably at the point indicated in Figure 14, as a consequence of imperfect polarization the beams reflected and transmitted by individual grids for frequencies above $\omega = 2c/s$ (a detailed analysis of this can be found in Ref. 1).

In Section 3.1, it is shown how calibration with thermal sources of known power spectral densities in effect determines the power-transmission functions for the several paths through the interferometer, allowing a frequency-dependent re-scaling of the outputs to compensate for the lost signal power. Detector noise (Section 2.9) will be similarly re-scaled in this calibration process, however, although it is not subject to the throughput loss; there is a consequent reduction of signal-to-noise ratio in the measured spectrum at frequencies where the power transmission through the interferometer is low. For this reason, an interferometer has a range of frequency over which it can perform well as a calibrated spectrum analyser but it will not give good signal-to-noise ratios at frequencies where the power-transmission functions are low. The wire spacings of grids used in an interferometer which is to record spectra up to ω_M, say, should therefore not be greater than about $2c/\omega_M$.

There is another non-ideality factor which limits the frequency range over which spectra can be determined with a particular interferometer. A wide-band detector is, of course, required for spectrum analysis but any one detector will have some band limitation. The electrical output from a wide-band detector is proportional to a weighted integral of the power spectral density of the incident signal beam over spectral frequency, the weighting function being the frequency-dependent responsivity of the detector system (including its collection optics—Section 2.6), $R(\omega)$ say. That is the detector output, D, will be

$$D = \int_{-\infty}^{+\infty} R(\omega) P_d(\omega)\, d\omega$$

where $P_d(\omega)$ is the PSD spectrum of the signal incident on the detector. For most types of wide-band detector, $R(\omega)$ will have a broad plateau between upper and lower cut-offs. The values of the cut-off frequencies vary with the type of detector.

The overall consequence is as if the PSD spectra of the incident beams had been reduced by the factor $R(\omega)$ and the detector had been ideal. The expressions above for the Fourier transforms of the interferograms in the two output ports must be multiplied by $R(\omega)$:

The FT for output port C3 gives:

$$R(\omega)[\,|H_{13}(\omega)|^2 P_1(\omega) - |H_{23}(\omega)|^2 P_2(\omega)]$$

The FT for output port C4 gives:

$$R(\omega)[-|H_{14}(\omega)|^2 P_1(\omega) + |H_{24}(\omega)|^2 P_2(\omega)]$$

The span of frequency over which a spectrum analysis can be made with an interferometric instrument will often be determined primarily by the cut-off frequencies of $R(\omega)$ for the detector installed in the instrument, rather than by a loss in the efficiency of beam division. The variation of $R(\omega)$ in the plateau range can be calibrated out by measurements using a source of known spectrum (Section 3.1) but, as indicated above for the loss of signal power due to beam-divider inefficiency, the loss of signal-to-noise ratio near and beyond the cut-off frequencies will band-limit the performance of the system.

2.4 POLARIMETRIC SPECTRUM ANALYSIS

In the treatment of the function of a polarization-division interferometer with ideal grids as a spectrum analyser in Section 2.2, the signal beams admitted at the two input ports of the interferometer came from separate sources and were superposed, with orthogonal directions of polarization, in a single beam emerging from the first grid in the divider (Figure 7).

If the first polariser in the divider is removed, however, the interferometer will admit a single signal beam and the interferometer will then process the two linearly polarized *components* of this beam as signals entering two input ports. Unlike beams from separate thermal sources, these orthogonally polarized component signal beams will not necessarily be statistically independent, however. In an arbitrarily polarized thermal signal beam there will be cross-correlations between the linearly polarized components. It will be shown below that a polarization-division interferometer can provide measures of such correlations. It will thereby fully determine the polarization properties of an arbitrarily polarized thermal signal beam.

The total field at a point in the beam in an output port of the interferometer, when the path difference in the interferometer is $c\tau$, is (from Section 2.1)

$$\tfrac{1}{2}[E_x(t+\tau) + E_y(t+\tau)] \pm \tfrac{1}{2}[E_x(t) - E_y(t)]$$

where $E_x(t)$ and $E_y(t)$ denote the time-dependent E-fields in the two input ports which are here identified as the linearly polarized components of the single input beam, in polarization directions (x and y) symmetrically disposed with respect to the wires of the remaining grid in the divider. The ± sign alternatives apply to the fields in the two output ports, respectively.

The detector in an output port will output a signal proportional to the square of this field averaged over time, i.e.

$$\{\langle[E_x(t)]^2\rangle + \langle[E_y(t)]^2\rangle\} \pm [\langle E_x(t)E_x(t+\tau)\rangle - \langle E_y(t)E_y(t+\tau)\rangle]$$
$$\pm [\langle E_x(t)E_y(t+\tau)\rangle - \langle E_y(t)E_x(t+\tau)\rangle]$$

where the angle brackets $\langle\ \rangle$ indicate averages over time. This function of τ is the interferogram.

The second term involves auto-correlation functions of the linearly polarized components whilst the third term involves cross-correlations between the components.

There are four correlation field terms here: $\langle E_x(t)E_x(t+\tau)\rangle$, $\langle E_y(t)E_y(t+\tau)\rangle$, $\langle E_x(t)E_y(t+\tau)\rangle$ and $\langle E_y(t)E_x(t+\tau)\rangle\}$. The spectra of all four would completely characterize a beam. It is customary, however, when specifying the polarization state of a beam, to use not the spectra of these terms but those of the following four independent linear combinations of these terms (known as the *Stokes parameters* [8]):

$$[\langle E_x(t)E_x(t+\tau)\rangle \pm \langle E_y(t)E_y(t+\tau)\rangle]$$
$$[\langle E_x(t)E_y(t+\tau)\rangle \pm \langle E_y(t)E_x(t+\tau)\rangle]$$

It can be seen that two of these combinations (those with the lower signs) are involved in the expression for the interferogram above and taking the Fourier transform of the interferogram will thus provide measures of the spectra of two of the four parameters required to characterize fully the incident beam.

It is important to see how the two can be separated in the Fourier transformation. The autocorrelation term

$$\langle E_x(t)E_x(t+\tau)\rangle - \langle E_y(t)E_y(t+\tau)\rangle]$$

is symmetrical about the origin $\tau = 0$ (Figure 15). That this is so can be confirmed by shifting the origin for t by τ, to obtain

$$[\langle E_x(t-\tau)E_x(t)\rangle - \langle E_y(t-\tau)E_y(t)\rangle]$$

(such a shift has no consequence for the averages over time indicated by the angle brackets $\langle\ \rangle$) and then noting that the substitution $\tau \longrightarrow -\tau$ re-orders but does not otherwise change the expression.

The cross-correlation term

$$[\langle E_x(t)E_y(t+\tau)\rangle - \langle E_y(t)E_x(t+\tau)\rangle]$$

on the other hand, is anti-symmetrical in τ (Figure 15); this can be confirmed by following the procedure above and noting that the original expression is

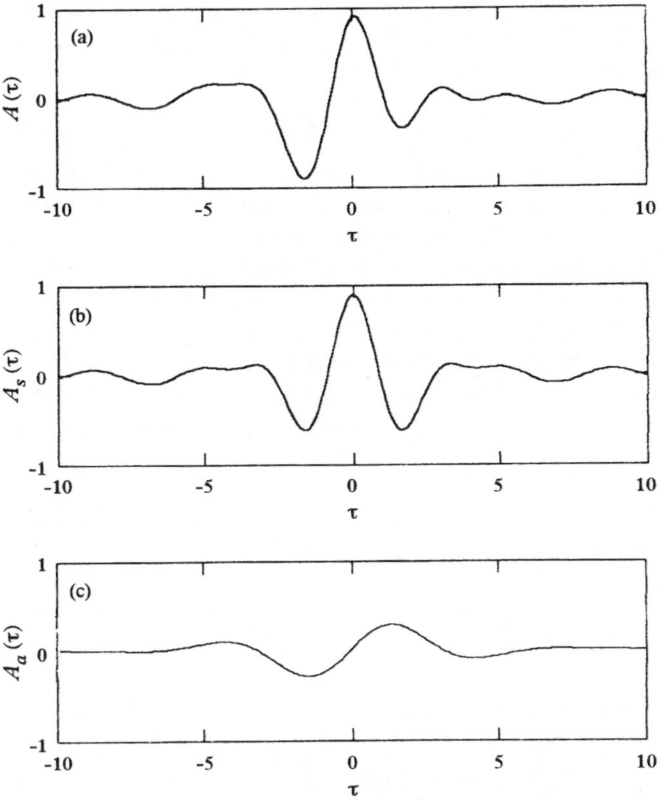

Figure 15 An interferogram produced by an interferometer configured for polarimetry, will be asymmetric about $\tau = 0$ to some degree. This is illustrated in (a) by an example, $A(\tau)$, for which $A(-\tau) \neq A(\tau)$. Any asymmetric function can be represented as a superposition of a symmetrical component, $A_s(\tau)$ say, $[A_s(-\tau) = A_s(\tau)]$, and an antisymmetric component $A_a(\tau)$ say, $[A_a(-\tau) = -A_a(\tau)]$, and this is illustrated in (b) and (c). The separation of a function into its symmetrical and antisymmetric components is unique

regained but with a change in sign. The consequence of this is that, on taking the Fourier transform of the interferogram (i.e. of the second and third terms together), the second term provides the real part and the third term the imaginary part of the transform; or, equivalently, the cosine transform of the interferogram gives the spectrum of the auto-correlation terms, and the sine transform gives the spectrum of the cross-correlation terms. The two terms are thus separable.

The spectrum of the auto-correlation terms is clearly the difference between the power spectral density spectra of the orthogonal linearly polarized components of the incident beam, say

$$P_x(\omega) - P_y(\omega)$$

The spectrum of the cross-correlation terms is a cross spectral density; a non-zero value at any frequency indicates correlated fluctuations of the two linearly polarized component fields that imply transient elliptical polarization at that frequecy; it provides a measure of the statistical mean value of the ellipticity.

To obtain a measure of another of the Stokes polarization parameters, the inteferometer should be rotated through $\pi/4$ around the line of incidence of the beam (or, equivalently, the beam optically rotated through $\pi/4$) and the instrument run again. The new components symmetrically disposed with respect to the grid wires are, in terms of the original components,

$$E'_x(t) = (1/\sqrt{2})[E_x(t) - E_y(t)] \qquad \text{and} \qquad E'_y(t) = (1/\sqrt{2})[E_x(t) + E_y(t)]$$

The powers in the output ports are given by the same expressions as before but with the new components replacing the original. The term symmetrical about $\tau = 0$ is

$$\pm[\langle E'_x(t)E'_x(t+\tau)\rangle - \langle E'_y(t)E'_y(t+\tau)\rangle]$$

in terms of the new components. It can be seen that, when Fourier (cosine) transformed, this term provides the difference between the PSD spectra of the components of the incident signal beam linearly polarized in the x' and y' directions, i.e. at $\pi/4$ to the original x and y directions. This same quantity expressed in terms of the original components is

$$\pm[\langle E_x(t)E_y(t+\tau)\rangle + \langle E_y(t)E_x(t+\tau)\rangle]$$

and can be identified as one of the Stokes parameters listed earlier, but not determined in the first measurement.

The term anti-symmetrical about $\tau = 0$ is

$$\pm[\langle E'_x(t)E'_y(t+\tau)\rangle - \langle E'_y(t)E'_x(t+\tau)\rangle]$$

in terms of the new components; in terms of the original components this is

$$\pm[\langle E_x(t)E_y(t+\tau)\rangle - \langle E_y(t)E_x(t+\tau)\rangle]$$

which is clearly identical with one of the Stokes parameters listed earlier and determined in the first measurement; it provides simply a re-measurement of this parameter. Since it has been identified above as a measure of the statistical mean value of ellipticity, its invariance on rotation of the incident beam is not surprising.

The remaining Stokes parameter is

$$\langle E_x(t)E_x(t+\tau)\rangle + \langle E_y(t)E_y(t+\tau)\rangle$$

the transform of which is clearly the sum of the spectra of the the two components of the incident beam, i.e. the spectrum of the total power in the beam. To measure that, the first polarizer must be restored to the divider and the powers in the two components separately determined (Section 3.1).

The spectra of the four Stokes parameters that fully characterize an arbitrarily polarized wide-band thermal signal beam [8] can thus be measured with a polarization-division interferometer.

2.5 THE LIMIT TO SPECTRAL RESOLUTION SET BY THE MAXIMUM PATH DIFFERENCE

An interferometer measures an auto-correlation function, $A(\tau)$, over a finite range of path difference only, from $\tau = -\tau_{max}$, through $\tau = 0$, to $\tau = +\tau_{max}$, where $c\tau_{max}$ is equal to the maximum accessible path difference, d_{max} (see Figure 16). A measured spectrum, $P_M(\omega)$ say, can be computed as a truncated Fourier transform:

$$P_M(\omega) = \int_{-\tau_{max}}^{+\tau_{max}} A(\tau) \exp - i\omega\tau \, dt$$

but $P_M(\omega)$ will not be precisely the true spectrum of the signal beam. The way in which it differs from the true spectrum can be seen as follows.

The truncated ACF can be represented as the product of the complete ACF with a top-hat function [Figure 17(a)] having the value unity for $|\tau| < \tau_{max}$ and the value zero outside that range; the truncated Fourier transform is clearly identical with the complete Fourier transform of that product. The convolution

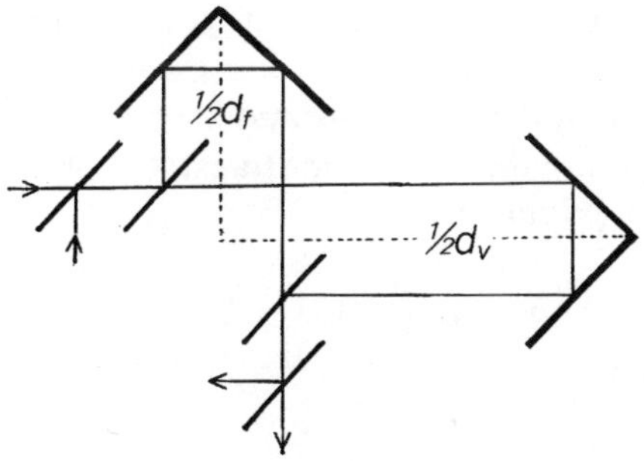

Figure 16 When recording an interferogram, one of the retroreflectors in the interferometer is moved along the direction of incident and emergent beams with the result that there is a changing value for the difference in length of the two paths in the interferometer, $d = d_f - d_v$, and in the associated time-delay, $\tau \equiv d/c$. Interferograms will be recorded over a bounded range of d, from $d = -d_{max}$ to $d = +d_{max}$, say, i.e. over a bounded range of time delay, from $-\tau_{max} \equiv -(d_{max}/c)$ to $+\tau_{max} \equiv +(d_{max}/c)$

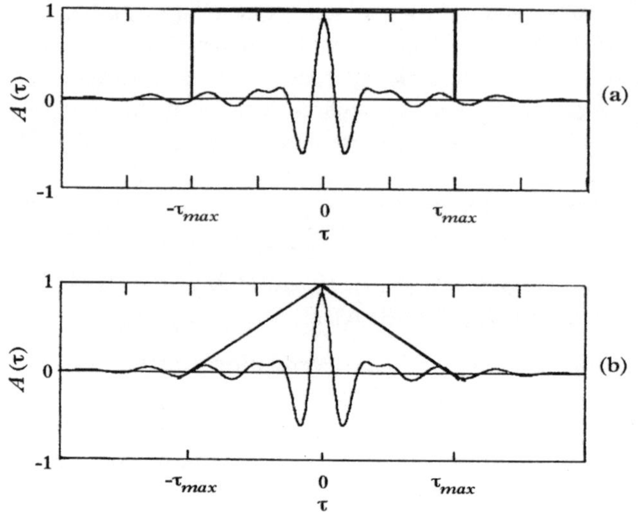

Figure 17 (a) An interferogram, $A(\tau)$, recorded between limits $\pm\tau_{max}$, can be represented as a product of the unlimited $A(\tau)$ with a top-hat function which takes the value 1 between $-\tau_{max}$ and $+\tau_{max}$ and the value 0 elsewhere, as illustrated here. (b) A linearly apodized interferogram is obtained by multiplying the limited $A(\tau)$ by the triangular function having value $\{(1 - |\tau)/\tau_{max})$ between $-\tau_{max}$ and $+\tau_{max}$ and the value 0 elsewhere, as illustrated here, and this is equivalent to the product of the unlimited $A(\tau)$ with the triangular function

theorem on Fourier transformation (which states that the Fourier transform of the product of any two functions is equal to the convolution of the Fourier transform of the first with the Fourier transform of the second) then leads directly to the conclusion that the measured spectrum, $P_M(\omega)$, is the convolution of the true spectrum with the Fourier transform of the top-hat function, i.e. with the sinc function (Figure 18)

$$R_{\tau\,max}(\omega) = 2\tau_{max}\frac{\sin\omega\tau_{max}}{\omega\tau_{max}}$$

That is,

$$P_M(\omega) = P(\omega) * R_{\tau_{max}}(\omega) \equiv \int_{-\infty}^{+\infty} P(\omega')R_{\tau_{max}}(\omega - \omega')\,d\omega'$$

where the $*$ denotes the process of convolution which is made explicit in the integral. The significance of the convolution integral is illustrated diagrammatically in Figure 19. Clearly, any detailed feature in the true spectrum having a frequency width smaller than the width of $R_{\tau_{max}}(\omega)$, i.e. π/τ_{max}, will not be

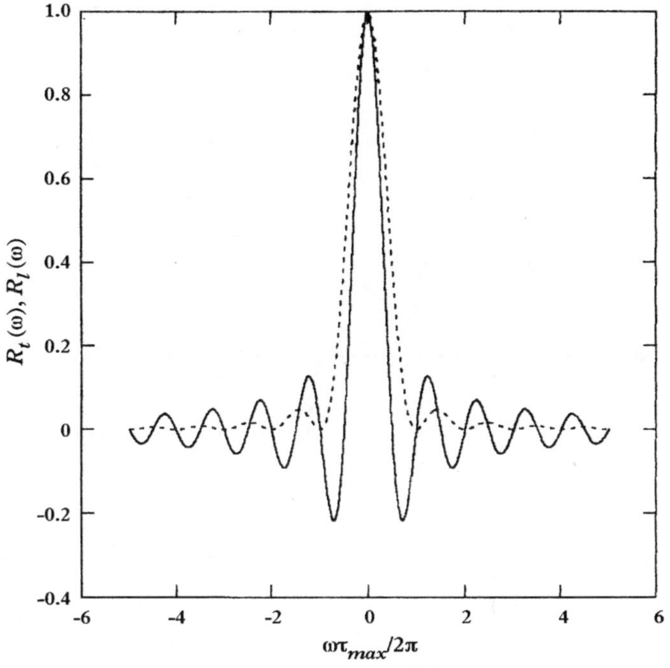

Figure 18 The function $R_t(\omega) = \sin \omega\tau_{max}/\omega\tau_{max}$ shown here by the full line is the instrument resolution function appropriate to an interferogram recorded from $-\tau_{max}$ to $+\tau_{max}$. The function $R_l(\omega) = [\sin(\omega\tau_{max}/2)/(\omega\tau_{max}/2)]^2$ shown by the broken line is the instrument resolution function appropriate to an interferogram linearly apodized between $-\tau_{max}$ and $+\tau_{max}$

resolved in the measured spectrum. $R_{\tau_{max}}(\omega)$ is for this reason known as the 'instrument resolution function'; the larger is τ_{max}, the narrower is the width of $R_{\tau_{max}}(\omega)$ and the higher the spectral resolution achieved.

The process known as apodization is sometimes used to reduce the chances of making a false inference that there are weak features close to a strong line in a measured spectrum when in fact only the subsidiary maxima of the sinc function associated with the strong line are involved. Apodization can best be illustrated by reference to a special case, namely linear apodization. In this the measured interferogram is linearly tapered to reach zero at $\tau = \pm\tau_{max}$ by multiplication with $(1 - |\tau|/\tau_{max})$ [Figure 17(b)]. The instrument resolution function to be included in the convolution is then the Fourier transform of $(1 - |\tau|/\tau_{max})$, rather than of the top-hat function, i.e. the square of the sinc function rather than the sinc function itself,

$$R_{\tau_{max}}(\omega) \equiv \tau_{max}\left[\frac{\sin\omega\tau_{max}/2}{\omega\tau_{max}/2}\right]^2$$

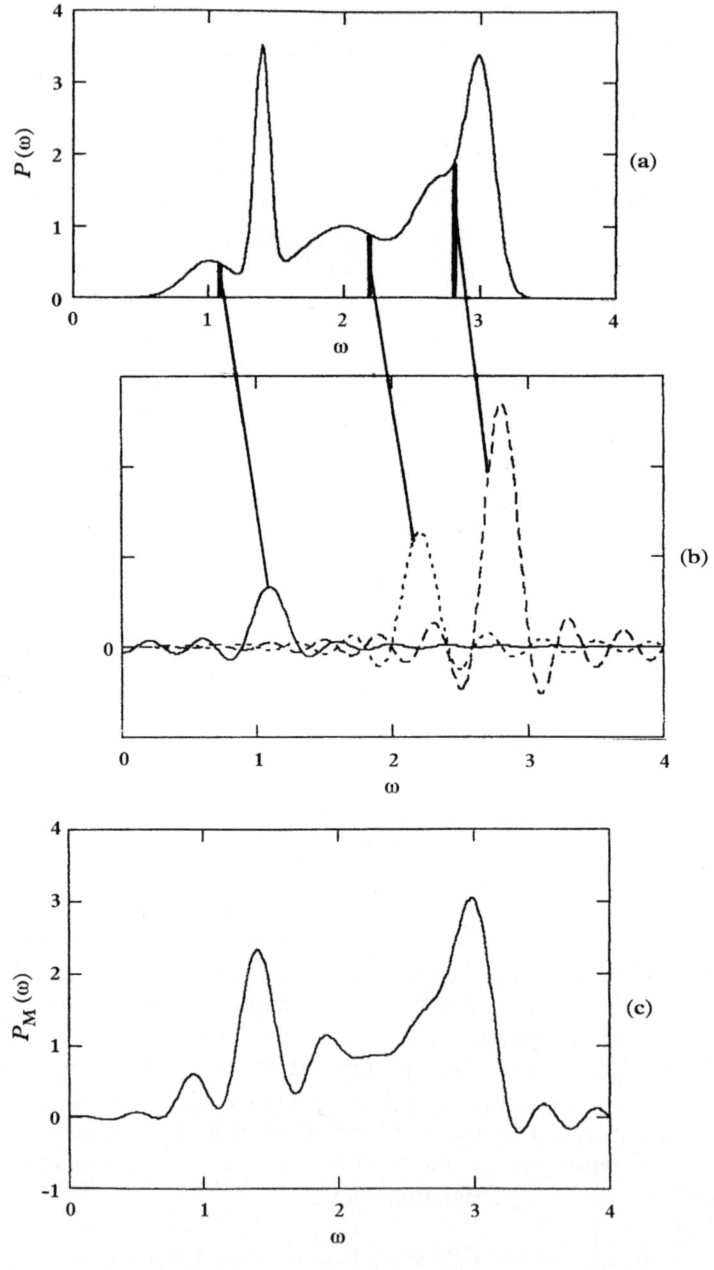

4
3
2
1
0
P (ω)
0
1
2
3
4
ω
(a)
0
0
1
2
3
4
ω
(b)
4
3
2
1
0
-1
P_M (ω)
0
1
2
3
4
ω
(c)

This is shown in Figure 18 and compared there with the unapodized resolution function; the subsidiary maxima are much reduced in amplitude but this is at the expense of a widened central peak. The consequences of apodisation for a measured spectrum are illustrated in Figure 20.

Other apodization functions can be selected but have the same general effect, a greater suppression of the subsidiary maxima being accompanied by a wider main peak.

Since the form of the instrument resolution function is known, it would be possible in principle computationally to de-convolve an unapodized measured spectrum to discover more precisely the true spectrum. For example, if the interferogram were known over a limited range, the derivatives of all orders at a point in the interferogram could in principle be determined, and a Taylor series expansion used to continue the interferogram to values of τ exceeding τ_{max}. That is a matter of principle, not of practice, however, because a real interferogram is recorded at discrete sampling points and each sampled point will be subject to measurement uncertainty or noise. Calculated higher derivatives would therefore be subject to large uncertainties and the extended interferogram would be subject to errors rising rapidly with the range of extension. It is unlikely that a significant improvement in resolution beyond that indicated by the instrument resolution function (π/τ_{max}) can be obtained even when the signal-to-noise ratio in the interferogram is extremely high.

Figure 19 The 'measured spectrum,' $P_M(\omega)$, of a signal beam is a superposition of sinc-function spectra each deriving from the signal power in a narrow frequency increment of the true spectrum of the beam, $P(\omega)$, as illustrated here. An arbitrary illustrative spectrum, $P(\omega)$, is shown in (a) and increments at three arbitrarily selected frequencies are indicated. The three sinc-function spectra contributed to $P_M(\omega)$ by these three increments are shown in (b). Each is centred at the frequency of the increment from which it derives and has a height scaled in proportion to the magnitude of $P(\omega)$ at that increment. The widths of the contributory sinc functions are all the same, and are determined by the range of time delay over which the interferogram is recorded (as in Figure 18). When sinc-function contributions from all contiguous increments in $P(\omega)$ are superposed in this way, the measured spectrum $P_M(\omega)$ shown in (c) results. Superposing sinc-function spectra in this way is a discrete convolution process. It will approximate well the result of an integral convolution so long as the width chosen for the increments is much less than the width of the sinc functions. In calculating the sinc functions shown in (b) and implied in (c), the range of time delay covered by the interferogram was arbitrarily set at $\pm\tau_{max} = \pm 10$, in units corresponding to those of ω in (a). It is clear in (c) that this value corresponds to barely resolving the widths of the lines in the spectrum. Had a larger value of τ_{max} been assumed, the resultant $P_M(\omega)$ would have more closely reproduced $P(\omega)$ because the contributory sinc-function spectra would have been narrower and more closely spaced

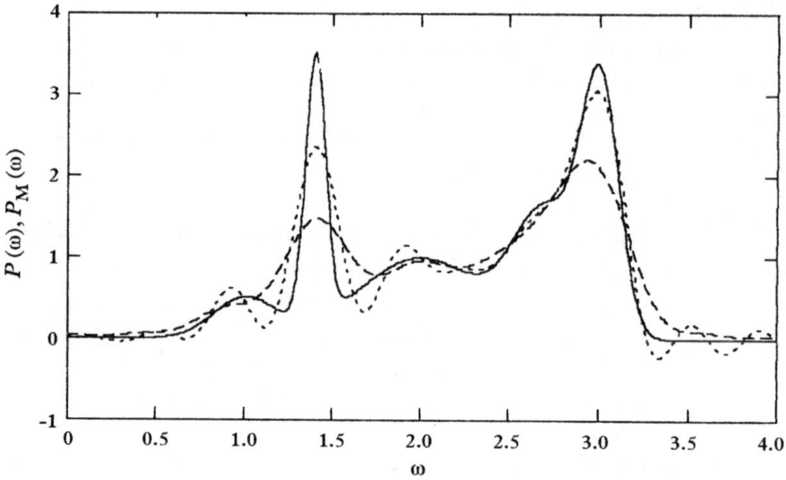

Figure 20 To illustrate the consequences for a measured spectrum of recording the interferogram over a limited range of time delay, with and without linear apodization. The full line is an arbitrary illustrative spectrum $P(\omega)$. The dotted line shows the spectrum obtained by transformation of an interferogram corresponding to the spectrum in (a) recorded over a limited range of τ, from $-\tau_{max} = -10$ to $+\tau_{max} = +10$. The broken line shows the spectrum obtained by transformation of the interferogram after linear apodization between $-\tau_{max} = -10$ and $+\tau_{max} = +10$. (See the comments on units and resolution in the caption to Figure 19)

2.6 THROUGHPUT AND RESOLUTION

It has been indicated in Sections 2.2 and 2.3 that a two-beam interferometer will measure the spectrum of a collimated signal beam. No real beam of finite width can be perfectly plane-parallel. A collimated beam will be a superposition of plane-parallel beams having a narrow spread of directions about the central line of propagation. That is, the signal beam will have a spatial spectrum in addition to a temporal spectrum. Both the spatial spectrum and the temporal spectrum of an incident beam will influence the distribution of total signal power in the interference pattern produced by an interferometer, but it would not be possible to separate the two in an analysis of a measured distribution. A two-beam interferometer can be used as a spectrum analyser (measuring the temporal spectrum of a signal beam) only when the beam's spatial spectrum is sufficiently band limited, i.e. when the beam is sufficiently well collimated. A quantitative criterion is required for the maximum acceptable spatial bandwidth because imposing a smaller bandwidth would result in an unnecessary reduction of signal power.

To find that criterion, note first that a collimated beam that entered an interferometer along an off-axis direction would suffer a larger path difference inside the interferometer than would an on-axis beam, by a factor $1/\cos\theta$, where θ is the off-axis angle. If no correction were made for this, a feature in the true spectrum at frequency $\omega = \omega_0$, say, would appear in the uncorrected measured spectrum at frequency $\omega = \omega_0/\cos\theta$. For an incident beam which is a superposition of plane-parallel beams all having the same temporal structure but with a continuous distribution of propagation directions between the axis and the off-axis angle $\theta = \theta_0$, the measured spectrum, $P_M(\omega)$, will be a convolution of the true spectrum, $P(\omega)$, with a resolution function $R_s(\omega)$, i.e.

$$P_M(\omega) = \int_{-\infty}^{+\infty} P(\omega - \omega') R_S(\omega')\, d\omega'$$

and the form of $R_s(\omega')$ as a function of ω' is essentially the same as the form of the directional distribution of power in the incident beam as a function of $(1 - \cos\theta)$ as illustrated in Figure 21, rising sharply from zero to a maximum value at $\omega' = 0$ and then falling to zero again at $\omega' = \omega(1/\cos\theta_0 - 1) \approx \omega(1 - \cos\theta_0)$. The width of this band, $\Delta\omega$, is

$$\Delta\omega = \omega(1 - \cos\theta_0) = \omega\Omega/2\pi$$

where Ω is the solid angle within which there are incident plane-parallel components in the signal beam as it passes through the interferometer. Thus, if the resolution required in a spectrum analysis is $\Delta\omega$ up to a frequency ω_{max}, not only must the maximum path difference, $c\tau_{max}$, be no less than $2\pi c/\Delta\omega$ (Section 2.5), but Ω must be no larger than

$$\Omega_R \equiv 2\pi(\Delta\omega/\omega_{max}) = 2\pi^2/(\omega_{max}\tau_{max}).$$

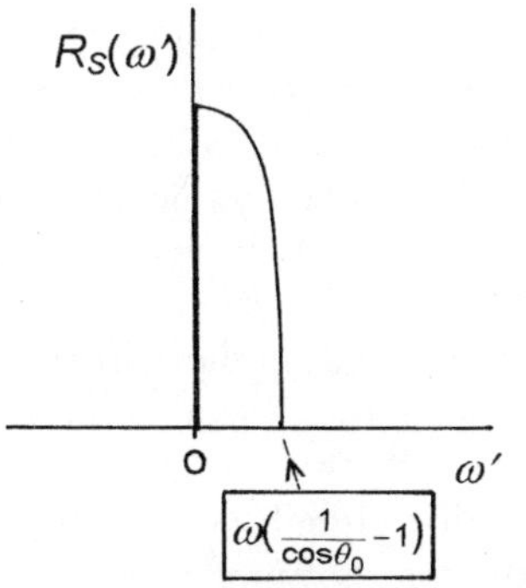

Figure 21 Resolution function for an imperfectly collimated signal beam

The detected signal power will be proportional to the area/solid angle product $A\Omega$, where A is the cross-sectional area of the signal beam in the interferometer (this product is variously called the throughput, the etendue or the light grasp). The larger is $A\Omega$, therefore, the shorter is the time that would be required to achieve a specified resolution and signal-to-noise ratio in a spectrum analysis. The distribution of signal power across a beam passing through an interferometer will not usually be one of uniformity out to a particular radius and then a sharp fall to zero, nor will the directional distribution be uniform and sharply bounded. The values of A and Ω should, in that case, strictly be defined as means, with weighting by the distribution. Sufficient quantitative information on the distributions will not usually be available for this, however, and it is customary to err in the direction of caution by using the upper limits for A and Ω defined by the edges of the apertures at lenses or reflectors, and at sources or detectors (see Figure 22).

When first designing an instrument, the value for A will be chosen after weighing the advantage of shorter measurement times against the demand on space, inconvenience in operation and the cost of a larger instrument.

The maximum throughput consistent with a specified resolution is obtained when $\Omega = \Omega_R$. To approach this condition usually requires input and output optical systems which match the source and detector to the interferometer. The input system is a lens or focusing reflector which collects signal power diverging from the source and produces a collimated beam; conservation of energy means that the area/solid angle product of the emergent beam must be the same as that of the collected beam [Figure 22(a)] but the balance between area and solid angle is changed by the lens or reflector to provide the Ω required in the interferometer. Similarly, the output system transforms the balance in the interferometer to that required at the detector [Figure 22(b)]. The intrinsic area/solid angle products of the sources and detectors involved in laboratory spectroscopy are usually larger than $A\Omega_R$ (except for measurements made at low resolving power with large interferometers). It is therefore usually necessary, in order that Ω should not exceed Ω_R, to limit the area at the source, or at the detector, with an aperture stop; the solid angles at the source and detector are limited by the apertures which define the beam diameter in the interferometer, often simply the edges of the lenses or reflectors in the input or output system (Figure 22). It is not necessary to impose an area limit at both source and detector since the value of area/solid angle product for the beam that passes through the interferometer to the detector is the smaller of the area/solid angle products in the input and output systems. The limiting aperture is usually placed at the source.

When signal power originates in a distant source that subtends a solid angle at the interferometer, Ω_S say, which is smaller than Ω_R, there would be a gain in signal power if the incoming signal were collected with a focusing antenna having a larger area than the interferometer, and the beam then recollimated to match the area of the interferometer—see Figure 23. It will be clear that the solid-angle, Ω, of the beam in the interferometer will then be larger than Ω_S by a factor equal to the ratio of the two areas, with a corresponding increase in the detected signal power; a limit to this ratio is Ω_R/Ω_S since a larger value would take Ω beyond Ω_R.

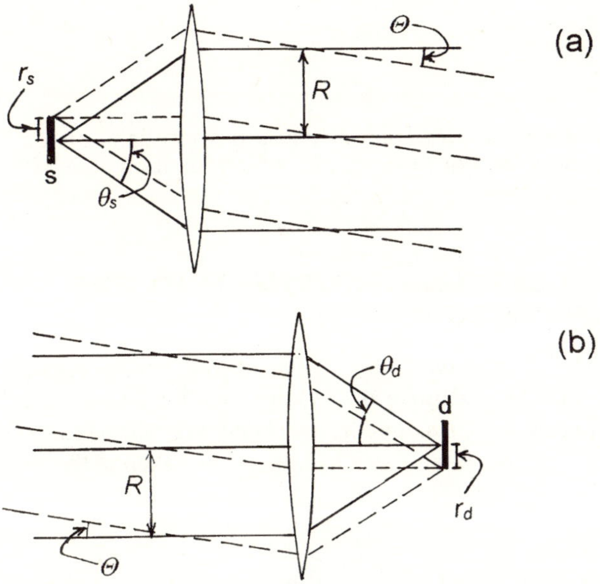

Figure 22 The source and/or the detector will be optically matched to an interferometer via a lens (as illustrated here) or equivalently via a focusing reflector. (a) Ray diagram for a beam emitted by a source, s, and collimated by a collecting lens pror to entering an interferometer. The collimated beam is a superposition of plane waves propagating over a range of directions about the central axis and out to an off-axis angle Θ determined by the radius of the source, r_s. The radius of the collimated beam is determined by the angle over which signal power is collected from the source, θ_s. (b) The quasi-collimated beam that is coupled to a detector, *d*, by a condensing lens. The angular spread, Θ, and radius, *R*, of the coupled beam are determined, respectively, by the radius of the sensitive area of the detector, r_d, and the angular range over which the detector collects and detects signal power, θ_d. The geometry of the ray paths in (a) indicates that the area–solid angle product of the beam collected from the source by the collimator is equal to that of the collimated beam, i.e. $a_s\omega_s = A\Omega$ (in which $a_s = \pi r_s^2$ is the area of the source, $\omega_s = \pi\theta_s^2$ is the solid angle over which signal power is collected from the source, and $A = \pi R^2$ is the area, and $\Omega = \pi\Theta^2$ the solid angle, of the collimated beam). Similarly, the geometry of the ray paths in (b) indicates equal area–solid angle products for the quasi-collimated beam that is selected by the condensing lens, and the beam that is admitted by the detector, i.e. $a_d\omega_d = A\Omega$ (where $a_d = \pi r_d^2$ is the area of the detector and $\omega_s = \pi\theta_s^2$ is the solid angle over which signal power is admitted by the detector)

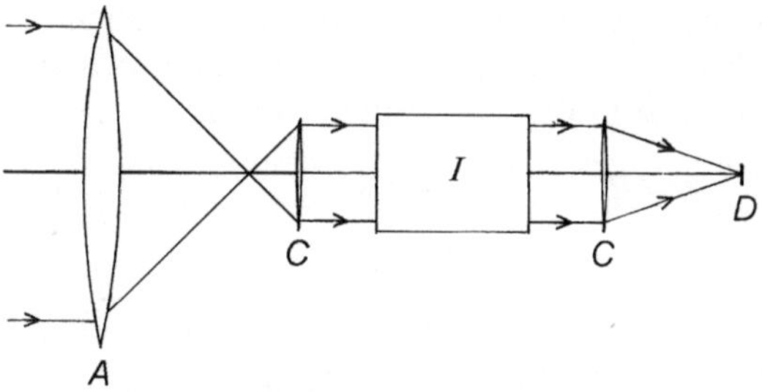

Figure 23 To illustrate an antenna with a primary lens or focusing reflector, A, and a secondary lens or reflector, C, which function as a beam compressor, as described in the text

2.7 DISCRETE SAMPLING AND DISCRETE FOURIER TRANSFORMATION

The interferogram produced by a two-beam interferometer is sampled discretely by the data acquisition system, with a constant interval of path difference between samples, and is recorded digitally. It is of finite length, being made up of an even number of samples, N, taken at the time delays

$$\tau = n\Delta\tau$$

where n denotes the integers $(-N/2+1)$, $(-N/2+2), \ldots, -1, 0, +1, \ldots, (N/2)$, and $\Delta\tau$ is the fixed increment in time delay. The overall time delay in the interferogram is thus $\tau_{\max} = N\Delta\tau$. We denote this discretely sampled interferogram $A(\tau = n\Delta\tau)$.

The integral Fourier transform in Section 2.2 cannot be used to determine a spectrum from a sampled interferogram; instead, the equivalent summation known as the discrete Fourier transform is required, i.e.

$$P(\omega = k \cdot 2\pi/\tau_{\max}) \equiv \sum_{n=-(N/2)+1}^{n=+(N/2)} A(\tau = n\Delta\tau) \exp\left(\frac{-i \cdot 2\pi nk}{N}\right)$$

where k denotes the integers 1 to $N/2$. The discrete Fourier transform, $P(\omega = k{\cdot}2\pi/\tau_{\max})$, is thus made up of $N/2$ data corresponding to discrete frequencies ranging from $\omega = 0$ to $\omega = \omega_m \equiv \pi/\Delta\tau$, with interval $\Delta\omega = 2\pi/\tau_{\max}$. Provided that the signal beam's true spectrum is band-limited, with no power above $\omega_m \equiv \pi/\Delta\tau$, this discrete Fourier transform can be identified as the discretely sampled PSD spectrum of the signal beam. If, however, the true spectrum of the beam did have power at frequencies above $\omega_m \equiv \pi/\Delta\tau$, that power would appear in the discrete Fourier transform folded back into the range $\omega = 0$ to ω_m as a contamination. This known as aliasing [6]. To avoid it, the sampling interval $\Delta\tau$ must be sufficiently short, so that $\omega_m \equiv \pi/\Delta\tau$ does exceed the largest frequency at which there is power in the true spectrum of the beam. If the spectrum of a beam is required only up to a specified frequency, ω_s say, and there is a wish to minimize the sampling rate, an anti-aliasing low-pass filter with cut-off frequency near $2\omega_s$ should placed in the signal beam (just prior to the detector is a common location); alternatively, a low-pass filter could be placed in the data acquisition system prior to digitization of the signal, with cut-off frequency near $2\omega_s(v/c)$, where v is the speed of movement of the reflector in the interferometer and c the speed of light. With one or other of these filters in place, the required sampling interval would be $\Delta\tau \leqslant \pi/2\omega_s$.

The interval in frequency between successive points in the discrete Fourier transform

$$\Delta\omega = 2\pi/N\Delta\tau = 2\pi/\tau_{max}$$

represents the limited spectral resolution that is the unavoidable consequence of the limited length of the interferogram, τ_{max}; it is equal to the resolution obtained with a continuous Fourier transform of a continuous interferogram of similar length (Section 2.5). This limit to resolution is not a consequence of discrete sampling; the sampling theorem of signal analysis [6] shows that there is an interpolation scheme which would allow the continuous transform of the continuous interferogram to be precisely constructed from the discrete transform (provided that the sampling interval for the interferogram is short enough to avoid aliasing). Although it might be more pleasing to see a continuous plot of a spectrum (and a simple interpolation scheme is often used when presenting spectra graphically), the underlying limit to the resolution that is inherent in the limited maximum path difference would remain.

Discrete Fourier transformation of a sampled interferogram is an exercise in numerical analysis, of course, and is directly undertaken by digital computation. There is a fast Fourier transform algorithm which dramatically reduces the number of elementary multiplications that have to be made in a discrete Fourier transform when N is a power of 2. Personal computers now have ample capacity and speed for the transformations involved in high-resolution Fourier transform spectral analysis and the associated data analyses.

2.8 ALIGNMENT AND SURFACE ACCURACY OF OPTICAL COMPONENTS

To achieve efficient interference, an interferometer must be well aligned and the surfaces of components must have high surface quality. The precision required in these matters is examined in this section.

With regard to the *surface quality* of the planar reflectors and grids between the divider and combiner, the rms deviations of the actual surfaces from true planes must be less than a small fraction of the shortest wavelength to which the interferometer is to be used. This is so because the signal beams which travel by the fixed and variable paths respectively will not interfere maximally, when superposed in the output, if their phase fronts are differently perturbed between divider and combiner.

The *precision of alignment* of the optical components in an interferometer must be such as will result in full interference between the four beams superposed in each output port (Section 2.2) synchronously across the full width of the beams. This requires that the axes of propagation of the beams nearly coincide; in deciding precisely how nearly, three aspects of misalignment must be considered—non-parallelism of the beams' axes of propagation, lateral displacements of parallel axes, and rotational displacements around coincident axes. (Only the first of these arises for the simplest of interferometer structures shown in Figure 10.)

Non-parallelism of the propagation axes

If the angle between the beams' axes is β, synchronization of the interference across the full width of the beams, D, would be obtained only for wavelengths greater than $\lambda' = 4D\beta$ (at which wavelength there would be just one interference fringe across the width of the beams). There would be a sharp loss of signal level in recorded spectra at wavelengths below λ'. In other words, the planar components in an interferometer must be aligned so that $\beta < \lambda_m/4D$, where λ_m is the shortest wavelength required in the spectra to be recorded. That includes the requirement that the angle between the faces of a roof-edge reflector (Figure 12) be within $\lambda_m/4D$ of $\pi/2$. For example, for $D = 30$ mm and $\lambda_m = 10$ μm, the required orientational precision is <0.1 mrad.

Lateral displacements of beams with parallel axes

Lateral displacement arises if the roof-lines or the lines of intersection of the two reflecting surfaces in each arm of the interferometer are not equally displaced from the centre-lines of their incident beams (Figure 24). Such displacements have consequences since the beams are not spatially coherent across their full widths because the sources from which they are derived (and/or the detectors by which they are detected) are not point-like. The beams are collimated but they have an intrinsic spread angle, δ say, equal to the angle subtended by the source/detector at the collimator/condenser. The transverse coherence length at a wavelength λ' will be equal to about λ'/δ. The lateral displacements of the beam axes from each other, d say, must not exceed this coherence length for the shortest wavelength to which spectra are to be recorded, λ_m say, i.e. $d < \lambda_m/\delta$.

Now δ must not exceed $(2\Delta\nu/\nu_m)^{1/2}$ where $\Delta\nu$ is the resolution required in the spectra to be recorded (Section 2.6). When only a low resolution (large $\Delta\nu$) is required, a large δ, and therefore a larger signal power, would be possible

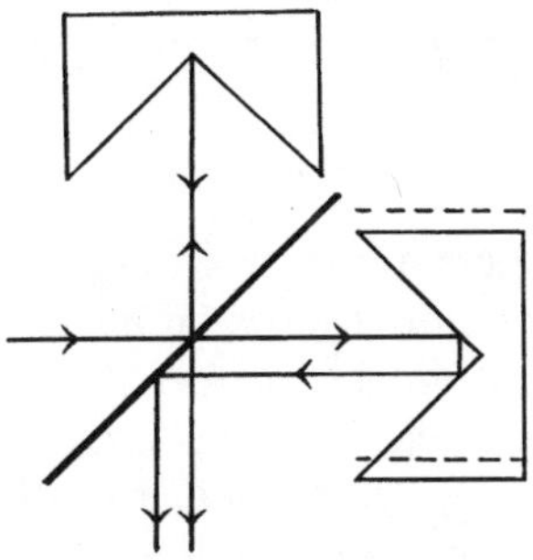

Figure 24 The roof-edge reflector on the right in this central unit of a Martin–Puplett interferometer is shown displaced downwards from its correctly aligned position (its edges would coincide with the broken lines when correctly aligned); as a consequence, there is a lateral displacement of one of the two emergent signal beams with respect to the other, as illustrated here by the relative diplacement of two rays, one in each emergent beam, which derive from a single incident ray

but the full signal-to-noise-ratio advantage of the larger signal power would accrue only if the alignment meets the correspondingly more demanding criterion above. The lateral displacement must be

$$d < (\lambda_m c/2\Delta\nu)^{1/2}$$

For example, for $\lambda_m = 10$ μm and $\Delta\sigma = \Delta\nu/c = 1\ \text{cm}^{-1}$, the maximum acceptable displacement is 0.2 mm.

The instrument will function if this requirement on alignment is not met but the signal power will not be fully interferometrically modulated; only that from the central region of the source within an angular radius equal to λ_m/d, where d is the achieved lateral displacement, would contribute to the modulation of the interferogram.

Rotational displacements of beams with coincident axes

The alignment could be such that the beams have a common axis but differ in their rotational displacements around it (this would be the case if the lines of intersection of the two reflecting surfaces in each arm were not precisely equally inclined to the plane of incidence). The difference in these angular displacements, α say, should be less than would correpond to a displacement at the perimeters of the beams equal to the transverse coherence length, i.e.

$$\alpha < (2\lambda_m c/\Delta\nu)^{1/2}/D$$

where D is the diameter of the beams. For example, for $\lambda_m = 10$μm, $\Delta\sigma = \Delta\nu/c = 1\ \text{cm}^{-1}$ and $D = 30$ mm, the rotational displacement should be < 8 mrad.

2.9 SIGNAL MODULATION AND DETECTION NOISE BANDWIDTHS

A wide-band detector is required for spectrum analysis. The electrical output from such a detector may have a non-zero value in the absence of an incident signal; it is the change in the output when an incident signal is switched on that provides the measure of the signal's power. The signal beam incident on the detector must therefore be modulated in some way. The character of the required modulation will be determined by the detector noise, i.e. the fluctuations in electrical ouput from the detector which occur whether or not there is a signal beam incident on it. Good design of a detector will minimize these noise fluctuations but could not eliminate them entirely because there are some fundamental origins of noise in the random thermal fluctuations of the distribution of energy of atomic motion in the materials of the detector. An interferogram recorded with an interferometer will include detector noise added to the true interferogram and the spectrum of the noise transfers to the Fourier transform of the interferogram; there is therefore an uncertainty error in each point of a measured spectrum due to detector noise. Minimizing this uncertainty is essentially a matter of maximizing the time over which the signal (with noise) is averaged, since the noise fluctuations have a zero mean value. An important feature of interferometric spectrometry is the fact that, since the signal power at all spectral frequencies falls on the detector throughout the measurement time, the effective averaging time for each point in a recorded spectrum is the total time spent acquiring the interferogram [3]. That is the case for each of the operational methods described below. There is advantage, nevertheless, in choosing for a given detector an optimum pattern of signal modulation since the spectra of the noise-fluctuations of most wide-band detectors are not flat against frequency (are not 'white').

For most wide-band detectors the output noise has the character of rapid fluctuations of lower amplitude superposed on slower variations of much larger amplitude; more precisely, the spectrum of the noise (i.e. the Fourier transform of the auto-correlation function of the fluctuations) is flat except that below some frequency, ν_N say, it rises strongly with decreasing frequency. For many wide-band detectors ν_N is in the range 10–100 Hz.

The several modulation procedures which have proved to be effective in avoiding the intrusion of the slower, larger fluctuations in detector output are described below.

Fast scanning and co-addition of interferograms or of spectra

An interferogram which is recorded as the path difference in an interferometer is changed at a constant rate is a function in time, $A(t)$ say. Provided that the origin for t is chosen to be the instant when the path difference passes through zero, the auto-correlation function of the signal field, $A(\tau)$, where τ is the delay time associated with the path difference, is obtained from the function $A(t)$ simply by the substitution $t \rightarrow \tau(c/v)$ (where v is the speed at which the path length is varied and c is the speed of light). The Fourier transform of $A(\tau)$ is the spectrum of the signal beam so the Fourier transform of $A(t)$ reproduces that spectrum but with frequencies scaled down by the factor v/c. (For example, a signal beam having a spectrum with a line at frequency 30 THz, wavelength 10 μm, when analysed by an interferometer in which the path varies at rate 1 mm s^{-1}, would give rise to an interferogram having a Fourier component at real-time frequency 100 Hz.) This means that a detector noise component at frequency ν' will be falsely read in the spectrum analysis as a signal component at spectral frequency $\nu'(c/v)$.

The increased levels of detector noise below the frequency ν_N will result in excessive noise in spectra recorded by the interferometer at spectral frequencies below $\nu = \nu_N(c/v)$. For this reason, there would be advantage in using a sufficiently high speed of reflector movement that this frequency lies below the lower bound of the range of frequency over which spectral analyses are required. That is what is meant by 'fast-scanning.'

A fast scan means a short integration time for each point in the interferogram, i.e. a short time spent averaging the detector output for each point, resulting in less noise suppression than would be given by a longer integration time. That would negate the noise advantage of fast scanning unless the data acquisition and control system allows repeated scans to be made, with acquisition of an interferogram in each scan, to take up the full available measurement time.

In this context, two data acquisition structures are to be considered: (a) co-adding the interferograms, followed by Fourier transformation of the result to give the recorded spectrum, and (b) Fourier transformation of each interferogram and co-addition of the resulting spectra.

The co-addition in each case improves the signal-to-noise ratio (the noise contributions vary from one record to the next, and are as often negative as positive at each point in a record) so that the full measurement time is effective in noise reduction (the noise-bandwidth is in both cases I/T, where T is the full measurement time).

The two procedures would yield the same final result if the acquisition process were perfect; the former would then be preferred on the grounds that it is less demanding of computer speed and capacity because many fewer Fourier transformations are required. However, if there are small errors in sampling, i.e. the values of path difference at which data sampling occurrs are not precisely the same for all interferograms, the consequential errors in the final spectrum would be less in the latter method than in the former.

The band pass of the signal acquisition system prior to the sampling of the interferogram must run from $\nu(v/c)$ to $\nu_u(v/c)$, where ν and ν_u are the lower and upper bounds of the range to be covered in a spectrum analysis. The Fourier transformation of the interferogram(s) is a filtering operation, however, and the co-addition of interferograms or spectra is an integration operation which has the effect of low-pass filtering; both contribute to a reduction in the overall noise bandwidth of the system. A detailed examination of this matter shows that the overall noise bandwidth for each point in a measured spectrum is $1/T$, where T is the total time spent in making a spectral analysis. This reflects the frequency multiplexing character of Fourier transform spectrometry, i.e. of the fact that all the incident signal power reaches one or other of the two output ports all the time a measurement is being made.

Slow scanning with signal modulation

The large-amplitude, slow, noise fluctuations in the detector output can be bypassed by frequent zero referencing. This involves repeated on–off switching of the signal beam, such as can be realized by placing a rotating bladed plate, or chopper wheel, in the signal beam. The switching or modulation frequency, ν_M say, should exceed the frequency, ν_N, below which the detector noise strongly rises. Modulating the signal in this way, while leaving the noise unmodulated, allows some discrimination of signal from noise; the output from the detector is passed through a filter with a pass-band centred on ν_M and of width $2\nu_u(v/c)$. The excess detector noise below ν_N is thus rejected. For a slow scan speed this is a narrow bandwidth and the filtering is best provided in the form of a lock-in amplifier (or phase-sensitive detector) with an output integrator to set the bandwidth.

The speed of the slow scan can be set so that the available measurement time is taken up in a single scan; alternatively, a more rapid scan might be used and the available measurement time taken up by repeated scans with co-addition of interferograms or spectra, as in the fast-scanning method. In any case, the final noise bandwidth for each point in the recorded spectrum is $1/T$, where T is the full measurement time.

The signal entering the interferometer from the second input port, being unmodulated, will not be detected. However, a thermal signal will be emitted from the blades in the chopper and will enter the interferometer modulated in anti-phase with the signal beam; the interferometer will consequently record a difference spectrum—the signal from the blades serves effectively as a signal in a second input port.

Slow scanning with output-port or input-port switching

The final component in a polarization-division interferometer is the wire grid at which the two output signals are separated, by reflection and transmission respectively. A rotation through $\pi/2$ of that grid around the direction of incidence of the signal beam interchanges the output beams—that which initially left by output port C3 will now leave by port C4, and *vice versa*. If such a rotation is made at regular intervals, and the detection system, placed in one of the ports, includes synchronous detection or lock-in amplification, it will be the difference of the interferograms in the two output beams that is recorded. Since the interferograms for the two output beams are complementary (see Figure 14), this difference interferogram has twice the amplitude of each interferogram separately; also, whereas the interferogram in each port is superposed on a DC level, the difference interferogram is made up of excursions about zero, as shown in Figure 25.

The modulation frequency at which this output-port switching occurs can be chosen to be above the frequency at which the detector in use exhibits excess

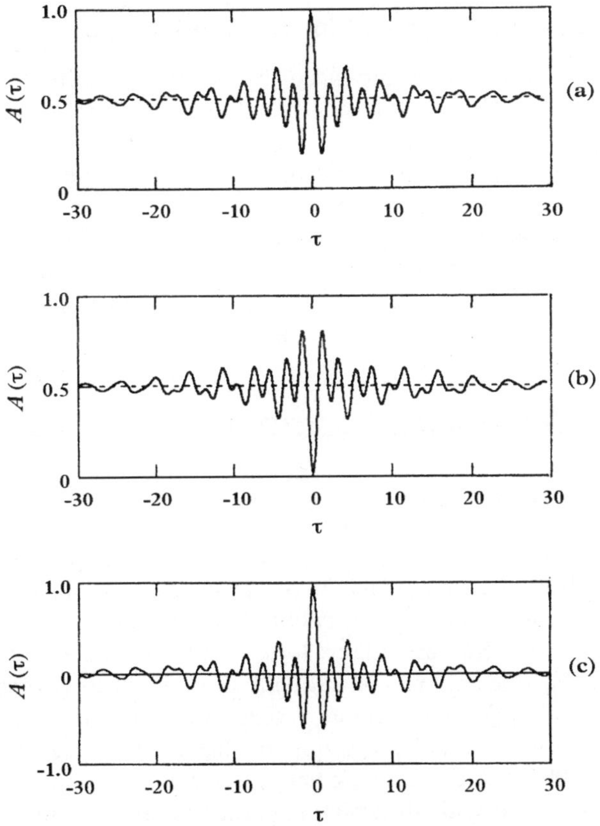

Figure 25 (a) and (b) show examples of complementary interferograms in the two output ports of an interferometer (as in Figure 14); (c) shows the difference interferogram that would be obtained recorded by switching between the output ports and detecting synchronously (phase-sensitive detection)

noise, to bring the same advantages as are gained with signal modulation as described above.

There is a further considerable advantage in output-port switching arising from the fact that the two output beams are complementary; as a result, any false signal-level fluctuations which are additive in the two beams (such as might be generated by vibrations of the grids) will be differenced in the switching and therefore suppressed in the difference interferogram.

The rotation of the grid need not be in rapid discrete $\pi/2$ shifts; continuous rotation is almost as effective and is more readily achieved. Moreover, this polarization switching can be effected at the input ports rather than at the output ports.

Slow scanning with differential (phase) modulation

Lastly, there is modulation produced by a small-amplitude reciprocating motion of the 'fixed' reflector additional to the slow scan of moveable reflector. The output from the detector as the slow scan develops is then essentially the derivative of the interferogram. The detection is at the frequency of the modulation, however, and the advantages of that are the same as those examined above for the other modulation techniques. The amplitude of the alternating motion of the reflector must be less than the shortest wavelength to be included in the spectral analysis and that results in reduced efficiency of detection (versus noise) at lower frequencies in the spectra.

3 FOURIER TRANSFORM SPECTROMETRY

3.1 SPECTRORADIOMETRIC MEASUREMENTS

A spectroradiometric measurement determines the power spectrum of a thermal signal radiated from a source under study. An interferometric spectrometer will provide such a measure but requires calibration both because the interferometer has frequency-dependent throughput losses and because the detector in the interferometer also does not respond equally throughout its operational band of frequency (Section 2.3). To calibrate a spectroradiometer, an absolute source of wide-band thermal radiation is required, that is, a source of known power spectral density. An isothermal black-body cavity provides this.

Inside a cavity lined with absorbing material and at uniform temperature, T_c say, there will be thermal radiation having a distribution of power versus frequency which is determined by the temperature T_c and fundamental constants only. If an aperture is introduced in the wall of such a cavity, it will radiate thermal power and, provided that its area is small compared with the internal surface area of the cavity, a measurement of T_c will fully determine the radiated power and its distribution in direction and in frequency. The aperture is a 'black-body' or 'Planckian' absolute source [9]. The radiated power is proportional to the area of the aperture and is isotropic in the forward half-space. The power per unit area, per unit solid angle and per unit frequency interval, in one linear polarization, i.e. the 'spectral radiance' of the source, $L_C(\nu)$, varies with frequency, ν, in accordance with the Planck equation:

$$L_C(\nu) = \frac{h\nu^3}{c^2} \frac{1}{\exp(h\nu/kT_c) - 1}$$

where h, c and k are Planck's constant, the velocity of light and Boltzmann's constant, respectively.

Two distinct radiometric situations must now be considered. In the first, the source under study is at a distance and is extended; in the second the source under study is nearby and localized.

Distant extended source

The radiation from the source under study is characterized by its spectral radiance, $L_S(\nu)$, i.e. the power radiated per unit area, per unit solid angle and per unit frequency interval in one linear polarization (that admitted by the first polarizer in the beam divider). $L_S(\nu)$ will vary over the surface of the source and the radiometer will measure an average over a patch on the source which is determined by the solid angle over which the radiometer will collect signal power, Ω_R say. The detector and its condensing optical system will determine Ω_R, and also the area at the input port to the radiometer over which signal power is admitted into the radiometer, A_R say, as described in Section 2.6. The signal power admitted by the radiometer is equal to $L_S(\nu)A_R\Omega_R$.

A calibration is made by placing a black-body cavity at an input port of the radiometer. The aperture of the cavity must exceed the size of the port, i.e. exceed A_R in area. The thermal signal power admitted by the radiometer from the cavity is then equal to $L_C(\nu)A_R\Omega_R$.

The values of A_R and Ω_R are mean values over non-uniform spatial and directional distributions of the beam admitted by the radiometer; there is, however, no need to know their values in a calibration procedure because the value of $A_R\Omega_R$ will be the same for the measurement with a source under study as for the measurements with calibration cavities.

An interferometer has two input ports and the signal entering the second port will also be involved, so the calibration procedure requires further elucidation. Suppose an extended source of spectral radiance $L_S(\omega)$ fills input port D1 and a black cavity of spectral radiance $L_A(\omega)$ fills input port D2 [see Figure 26(a)]; if $S_S(\omega)$ denotes the Fourier transform of the interferogram in output port C3,

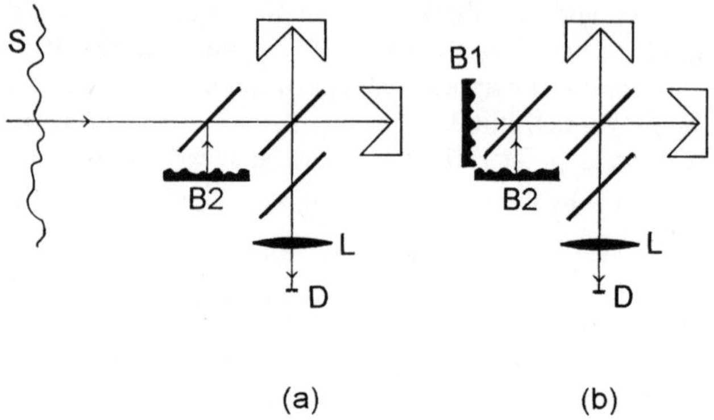

Figure 26 The two stages in a calibrated measurement of the spectral radiance of a distant extended source. In (a) the interferometer is directed at a distant extended source, S, and has a black cavity, B2, filling its second input port (the cavity is illustrated diagrammatically here). In (b) a second black cavity, B1, is interposed to fill the first input port

the expressions established in Section 2.3 indicate that

$$\frac{S_S(\omega)}{R(\omega) A_R \Omega_R} = |H_{13}(\omega)|^2 L_S(\omega) - |H_{23}(\omega)|^2 L_A(\omega)$$

It was remarked in Section 2.3, however, that there should be an additional term to allow for thermal emission from (non-ideal) grids and other ambient-temperature objects. This term is small but could be significant in a precision spectroradiometric measurement; it can be established by a general argument as follows. If the sources in the two input ports were both black cavities at the ambient temperature of the instrument and surroundings, the system would be itself a black cavity and sweeping the moveable reflector could not result in any modulation of the power at either output port. The expression above, with $L_S(\omega) = L_A(\omega) = L_a(\omega)$, where the subscript a indicates ambient temperature, does *not* give zero when the power transmission functions linking the two input ports with the output port, $|H_{13}(\omega)|^2$ and $|H_{23}(\omega)|^2$, are not equal. The following correction to the expression above is therefore required, to include the internal thermal emissions:

$$\begin{aligned}\frac{S_S(\omega)}{R(\omega) A_R \Omega_R} &= [|H_{13}(\omega)|^2 L_S(\omega) - |H_{23}(\omega)|^2 L_A(\omega)] - [|H_{13}(\omega)|^2 L_a(\omega) - |H_{23}(\omega)|^2 L_a(\omega)] \\ &= |H_{13}(\omega)|^2 [L_S(\omega) - L_a(\omega)] - |H_{23}(\omega)|^2 [L_A(\omega) - L_a(\omega)]\end{aligned}$$

If the second input port of the radiometer is covered by a black cavity at the instrument ambient temperature [Figure 26(b)] so that $L_A(\omega) = L_a(\omega)$, this becomes

$$\frac{S_S(\omega)}{R(\omega)A_R\Omega_R} = |H_{13}(\omega)|^2 [L_S(\omega) - L_a(\omega)]$$

i.e.

$$S_S(\omega) = K(\omega)[L_S(\omega) - L_a(\omega)]$$

where the response factor, $K(\omega)$, is $R(\omega)A_R\Omega_R|H_{13}(\omega)|^2$. (It is important that this ambient-temperature cavity be highly black if the detector in use is cryogenic.)

To effect a calibration, i.e. to determine $K(\omega)$, a black cavity, at measured temperature T_c say, is placed to cover the first input port [Figure 26(b)]. The Fourier transform of the resultant interferogram, $S_C(\omega)$ say, then determines $K(\omega)$:

$$K(\omega) = \frac{S_C(\omega)}{L_C(\omega) - L_a(\omega)}$$

where the spectral radiances for the temperatures T_c and T_a are known from the Planck equation above.

If now the first port of the radiometer is uncovered and directed towards a source under study [Figure 26(a)], and the Fourier transform of the resultant interferogram is $S_S(\omega)$, the spectral radiance of the source is given by

$$L_S(\omega) = \frac{S_S(\omega)}{K(\omega)} + L_a(\omega)$$

There is an operationally more convenient calibration procedure in which, except during an initial check procedure, the calibrating black cavities are placed in the second port and the first port remains uncovered, and directed towards the source under study, throughout. Moreover, the two calibrating black cavities, one at ambient teperature, T_a, and the other at a raised (or lowered) temperature, T_c, can be continuously in place if the initial polarizer can be rotated to select an input beam from one or the other (Figure 27). The procedure is to record an interferogram with the polarizer switched to the cavity at temperature T_a, and then to record a second interferogram with the polarizer switched to the cavity at T_c; in both measurements the source under study is in the first input port. If $S_1(\omega)$ and $S_2(\omega)$ denote the Fourier transforms of the two interferograms, the argument as before leads directly to the following determination of the spectral radiance of the source under study:

$$L_S(\omega) = \frac{r(\omega)S_1(\omega)}{S_1(\omega) - S_2(\omega)} [L_C(\omega) - L_a(\omega)] + L_a(\omega)$$

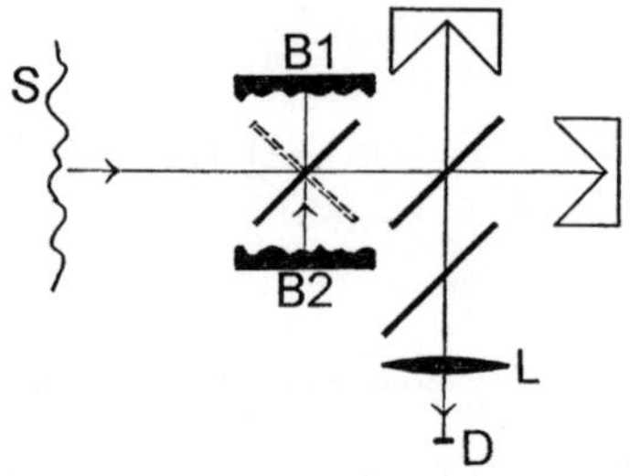

Figure 27 An alternative two-stage procedure for calibrated spectral radiance measurements. Two black-cavity sources, B1 and B2, and the distant source under study, S, are continuously in place as shown; the two measurements are made with the input polarizer in each of the two positions shown (one indicated by full lines and the other by broken lines)

where

$$r(\omega) = \frac{|H_{23}(\omega)|^2}{|H_{13}(\omega)|^2}$$

Determination of the value of $r(\omega)$ is the purpose of an initial procedure. Since this is a matter of the ideality of the grids, rather than of the detector's $R(\omega)$, it would not be necessary to make this determination frequently. The measurement required is to record an interferogram with the polarizer switched to the black cavity at temperature T_c and the first input port covered with the black cavity at ambient temperature T_a, and then to interchange these cavities and take a further interferogram. If $S_3(\omega)$ and $S_4(\omega)$ denote the Fourier transforms of these interferograms, the argument as before leads to

$$r(\omega) = \frac{S_3(\omega)}{S_4(\omega)}$$

Over the range of frequencies for which $r(\omega) \rightarrow 1$ the equation for $L_S(\omega)$ above simplifies to

$$L_S(\omega) = \frac{S_1(\omega)}{S_1(\omega) - S_2(\omega)} L_C(\omega) - \frac{S_2(\omega)}{S_1(\omega) - S_2(\omega)} L_a(\omega)$$

Losses of signal power in the interferometer can thus be calibrated out. This does not mean, however, that good performance can be obtained with an interferometric radiometer even over ranges of frequency where the power transmission or the detector responsivity is low. In the calibration process the spectrum of the output beams is subject to a frequency-dependent re-scaling to compensate for the lost signal power; however, detector noise (Section 2.9) is inevitably similarly re-scaled in this process (although it was not subject to a throughput loss), and there is a consequent reduction of signal-to-noise ratio in the measured spectrum at frequencies where the power transmission through the interferometer and/or the responsivity of the detector are low. Such deterioration of signal-to-noise ratio will set boundaries in frequency to the effective performance of an interferometric spectrum analyser.

Localized nearby source

A radiometric measurement involving a signal beam from a localized nearby source has a different character. What is required is a measure of the spectrum of the power carried by the beam rather than a determination of spectral radiance. The source will have a known shape and size, and the beam will diverge from it within a known solid angle (Figure 28). The collimator, condenser and detector of the radiometer must have dimensions that will focus all the output beam's power within the sensitive area of the detector. As described in Section 2.6, this means that the source, not the detetector, limits the throughput. (This is the converse of the requirement when a distant extended source is under study). However, although the sensitive area of the detector should for this reason exceed the area of the images of the source focused upon it, it should not greatly exceed

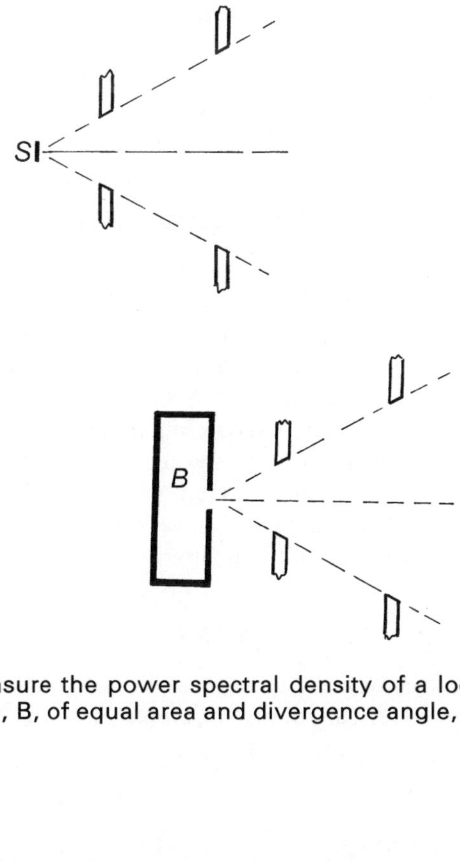

Figure 28 To measure the power spectral density of a localized source, S, a black-cavity source, B, of equal area and divergence angle, is required

it because that would imply that thermal signal power from surfaces outside the aperture of the source under study would be included in the recorded spectra.

To effect calibration, a black cavity with an aperture similar to that of the source under study and with aperture stops to define a solid angle similar to that of the source under study is required (Figure 28); all the signal power from such a calibrating source will be focused within the sensitive area of the detector. The power spectral density of the calibrating beam is the spectral radiance given by the Planck equation above multiplied by the aperture area and by the solid angle defined by the aperture stops.

The calibration is then a matter of comparing the spectra obtained with the source under study and with the calibration source, including corrections for the signal power collected in the second input port from its cover at ambient temperature and in the first input port from outside the aperture of source under study (these corrections involve the throughput as limited by the detector size).

It would be unusual for a source under study to have the uniform radiance over the aperture and the isotropic directional distribution within the solid angle defined by the stops, shown by a black-cavity source. The distribution of signal power within the detector's sensitive area will therefore differ for the calibration source and the source under study. If the response of a detector shows some dependence on the detailed distribution of incident signal power within its sensitive area, there will be consequential measurement error in the signal power determination.

It will be clear that absolute measurements of the power spectra of weak signal beams over wide spectral bands cannot be made as precisely as measurements of the radiance of extended sources.

Polarimetric measurements of a collimated signal beam

It has been shown in Section 2.4 that removal of the first grid in the beam divider enables determinations of the spectra of the polarization parameters of an incident beam to be made using an interferometer with ideal grids. When an analysis is carried through for an interferometer with non-ideal grids, as an extension of the procedures followed in Section 2.3, it is found that a clear separation of the spectra of the polarization parameters will be obtained at frequencies at which the grids are non-ideal, provided that the Fourier transform of the difference in the interferograms in the two output ports is taken [10] and the calibration procedures required for such polarimetric measurements are the same as those indicated above.

3.2 SPECTROSCOPIC MEASUREMENTS

In a spectroscopic measurement, the objective is to measure the frequency-dependent complex transfer function of a sample of a material in the form of a parallel-sided slab, in transmission, $H_T(\omega)$, or in reflection, $H_R(\omega)$, A transfer function $H(\omega)$ is defined by reference to the change in amplitude ($A_i \rightarrow A_e$, say) and shift in phase ($\omega t \rightarrow \omega t + \theta$, say) of a collimated coherent signal beam of angular frequency ω on transmission/reflection at the sample; the modulus of $H(\omega)$ is the ratio A_e/A_i and the phase angle of $H(\omega)$ is θ. The squared modulus, $|H(\omega)|^2$, is the power transmittance/reflectance.

In a spectroscopic measurement a wide-band thermal source is required (Section 1.1) to cover, with as much signal power as possible, the range of frequency over which reflection or transmission spectra are to be determined. That points to a high temperature for the source, e.g. gas in a high-pressure discharge tube or an electrically heated ceramic rod. The source, with collimator, is placed in the first input port of the interferometer and the second port is covered by a black cavity at ambient temperature; the measurement situation is thus similar to that considered in Section 3.1 for spectroradiometric studies of extended thermal sources, with the addition of provision for placing the sample. The location of the sample depends on whether the objective is to measure the complex transfer function $H(\omega)$ or the power transmittance/reflectance, i.e. $|H(\omega)|^2$. Both cases will be considered below.

Power transmission and reflection spectra

The sample in this case is between the interferometer and the detector/condenser, for transmission or reflection (Figure 29). An interferogram is recorded with the sample in place, and another with the sample removed (transmission) or replaced by a metallic reflector (reflection). The effect of the sample's power transfer is essentially similar to that of the frequency-

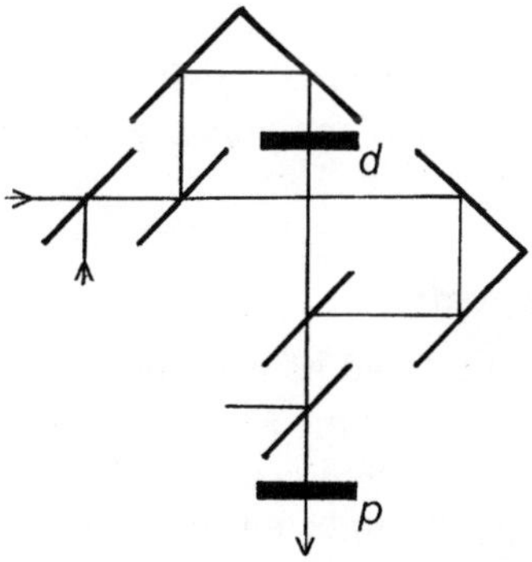

Figure 29 In using an interferometer to measure the power-transmission spectrum of a sample of a material in the form of a slab, the sample will be placed at the location indicated by p. To measure the amplitude and phase of the transmission spectrum of the sample, it will be placed in one arm of the interferometer, in position *d*

dependent responsivity of the detector so the equation in Section 3.1 leads directly to the following equations for the Fourier transform of the interferogram in output port C3, with sample, $S_M(\omega)$ say, and that without sample, $S_B(\omega)$:

$$\frac{S_M(\omega)}{|H(\omega)|^2 R(\omega) A_R \Omega_R} = |H_{13}(\omega)|^2 [L_S(\omega) - L_a(\omega)]$$

and

$$\frac{S_B(\omega)}{R(\omega) A_R \Omega_R} = |H_{13}(\omega)|^2 [L_S(\omega) - L_a(\omega)]$$

where $L_S(\omega)$ and $L_a(\omega)$ are the spectral radiances of the wide-band source and of the cavity at ambient temperature, respectively, $R(\omega)$ and $A_R\Omega_R$ are the responsivity and the throughput of the detector system, respectively, and $|H_{13}(\omega)|^2$ is the power transmission function for each of the two paths linking input port D1 and output port C3. The ratio of these equations gives

$$|H(\omega)|^2 = \frac{S_M(\omega)}{S_B(\omega)}$$

i.e. a straightforward ratio of the spectra obtained with and without the sample in the beam will give the power transmittance or reflectance of the sample.

The variations with frequency of the radiance of the source, of the reponsivity of the detector and of signal power transmission in the interferometer are thus calibrated out in taking this simple ratio. In this calibration process, however, detector noise (Section 2.9) is also re-scaled and there is a consequent reduction of signal-to-noise ratio in the measured spectrum at frequencies where any or all of these factors are low. Such deterioration of the signal-to-noise ratio will set boundaries on the frequency range over which the interferometric spectrometer will provide a good signal-to-noise ratio in the measured spectra.

Amplitude and phase spectra

By placing the sample in the fixed path inside the interferometer the complex transfer function for transmission, $H_T(\omega)$, can be determined, i.e. both amplitude and phase spectra. This is commonly known as dispersive Fourier transform spectroscopy [11]. It can be understood as follows.

In the absence of the sample in the fixed path, the Fourier transform of the interferogram in output port C3, $S_B(\omega)$ say, is (Section 3.1)

$$\frac{S_B(\omega)}{R(\omega) A_R \Omega_R} = |H_{13}(\omega)|^2 [L_S(\omega) - L_a(\omega)]$$

where $L_S(\omega)$ and $L_a(\omega)$ are the spectral radiances of the wide-band source and of the cavity at ambient temperature in the second input port, respectively, $R(\omega)$ and $A_R\Omega_R$ are the responsivity and the throughput of the detector system, respectively, and $|H_{13}(\omega)|^2$ is the power transmission function for each of the two paths linking input port D1 and output port C3.

When a sample is in place in the fixed path, the two paths (fixed and variable) are no longer identical apart from the difference in path lengths. The resultant interferogram is no longer an auto-correlation function; it is a cross-correlation function and its Fourier transform is a cross-spectral density, $S_M(\omega)$ say, given by (Section 1.2)

$$\frac{S_M(\omega)}{R(\omega)A_R\Omega_R} = H_T{}^*(\omega)|H_{13}(\omega)|^2 [L_S(\omega) - L_a(\omega)]$$

where the asterisk denotes the complex conjugate. $S_M(\omega)$ will be complex because it is the Fourier transform of a cross-correlation function (the interferogram) which will not be symetrical about $\tau = 0$.

A straightforward ratio of the two Fourier transforms thus provides a measure of the complex $H_T(\omega)$:

$$H_T(\omega) = \frac{S_M(\omega)^*}{S_B(\omega)}$$

A reflection transfer function can be determined in essentially the same way, although with some elaboration of the optical configuration in the arm of the interferometer containing the sample.

As remarked previously in this section, although the variation with frequency of the radiance of the source, of the reponsivity of the detector and of the signal power transmission in the interferometer are calibrated out in taking this simple ratio, the detector noise (Section 2.9) is similarly re-scaled in the calibration process and there is a consequent reduction of signal-to-noise ratio in the measured spectrum at frequencies where any or all of these factors are low. Such deterioration of signal-to-noise ratio will set boundaries on the frequency range over which the interferometric spectrometer will provide good signal-to-noise ratio in measured spectra.

4 REFERENCES

1. D.H. Martin and E. Puplett, *Infrared Phys.*, **10**, 105 (1970); D.H. Martin, in *Infrared Detection Techniques for Space Research*, Eds V. Manno and J. Ring, pp. 267–280, Reidel, Dordrecht (1972); D.H. Martin, in *Infrared and Millimeter Waves*, Ed. K.J. Button, Vol. 6, Chapt. 2, pp. 65–148, Academic Press, London (1982).

2. B. Carli and V. Natale, *Appl. Opt.*, **18**, 3954 (1979); J. Chamberlain, *The Principles of Interferometric Spectroscopy*, Wiley, New York (1979).
3. L. Genzel and K. Sakai, *J.Opt.Soc.Am.*, **67**, 871 (1977); P.R. Griffiths *Appl.Opt.*, **17**, 1315 (1978); *Science*, **222**, 297 (1983).
4. D.K Lambert and P.L. Richards, *Appl. Opt.*, **17**, 1595 (1978); W.G. Chambers, C.L. Mok and T.J. Parker, *J. Phys. D*, **13**, 515 (1980); B. Carli and F. Mencaraglia, *Int. J. Infrared Millimeter Waves*, **2**, 87 and 1045 (1981).
5. J.W. Goodman, *Statistical Optics*, pp. 68–81, 120–136, Wiley, New York (1985).
6. R. Bracewell, *The Fourier Transform and its Applications*, (Chapt. 9), McGraw-Hill, New York (1978); P.A. Lynn, *An Introduction to the Analysis and Processing of Signals*, Chapt. 5. and 7, Macmillan Education, London (1989); A. Papoulis, *Signal Analysis*, McGraw-Hill, New York (1977).
7. J.A. Beunen, A.E. Costley, *et al. J. Opt. Soc. Am.*, **71**, 184 (1981).
8. M. Born and E. Wolf, *Principles of Optics*, Sect. 10.8, Pergamon Press, Oxford (1980).
9. F. Grum and R.J. Becherer, *Optical Radiation Measurements; Radiometry*, Chapt. 4, Academic Press, London (1979); F.O. Bartell and W.L. Wolfe, *Infrared Phys.*, **16**, 13 (1976); J.M. Palmer, *Appl. Opt.*, **18**, 758 (1979).
10. D.H. Martin, in *Infrared and Millimeter Waves*, Ed. K.J. Button, Vol. 6, Chapt. 2, Sect. IIIA.2, Academic Press, London (1982).
11. J.R. Birch and T.J. Parker in *Infrared and Millimeter Waves*, Ed. K.J. Button, Vol. 2, Chapt. 3, pp. 137–271, Academic Press, London (1979).

2

Far-Infrared Absolute Spectrophotometer (FIRAS) Measures the Big Bang

JOHN C. MATHER
Code 680, NASA Goddard Space Flight Center, Greenbelt, MD 20771, USA

CONTENTS

Principles and Applications of Polarization-Division Interferometry
Edited by P. L. Polavarapu Published 1998 John Wiley & Sons Ltd

1 INTRODUCTION

According to Big Bang theory, the universe contains relic radiation left over from an early, optically thick, thermal stage in its history. This radiation was first detected by Penzias and Wilson [1] and since then, many investigators have attempted to determine the spectrum of the cosmic microwave background (CMBR) at far-infrared and radio frequencies. Its exact shape holds important clues to the physics of the universe during the first million years after the Big Bang.

Prior to 1989, the best constraints on the spectrum of the CMBR indicated a Planck-like shape with a peak near 2 mm (5 cm^{-1}, 150 GHz) and a temperature near 3 K. However, the measurements were sufficiently crude that a pure Planck distribution was not the only one consistent with the available information. In 1987, Paul Richards and Andrew Lange at the University of California at Berkeley, in collaboration with Toshiro Matsumoto and co-workers at Nagoya University, launched a rocket-borne filter photometer that measured the CMBR intensity at wavelengths not observable from the ground. Although several points did lie on the 3 K Planck curve, there appeared to be excess emission present on the short-wavelength side of the peak. The magnitude of the measured deviations from a Planck distribution, if independently verified, would have indicated some unexpected source of energy comparable to the total energy in the cosmic fireball itself. This in itself would have been a major change in observational cosmology nearly equal to the detection of the CMBR itself.

The FIRAS instrument, described by Mather, Fixsen and Shafer [2] and in the FIRAS Explanatory Supplement [3], is a dual-beam, Martin–Puplett interferometer specifically designed to be sensitive to deviations in the CMBR from a Planck spectrum smaller than 0.1%, more than two orders of magnitude better than ground- and balloon-based observations. The results from FIRAS [4–7] would soon show that this instrument had exceeded this ambitious goal, and confirmed that the CMBR spectrum was a virtually perfect blackbody as shown in Figure 1.

The use of a two-beam design to determine the detailed spectral properties of a radiation source was pioneered by Fellgett [9], Jacquinot [10] and Gebbie [11]. A number of technological modifications to their basic design have occurred over the years to improve its performance in the far-infrared and sub-millimeter regions (e.g. by Peck [12] and Hopkins [13]). The designs initially utilized a thin-film, dielectric beamsplitter, or a lamellar beamsplitter to divide the incoming beam, but these had throughput efficiencies that varied significantly with frequency, particularly in the infrared and sub-millimeter

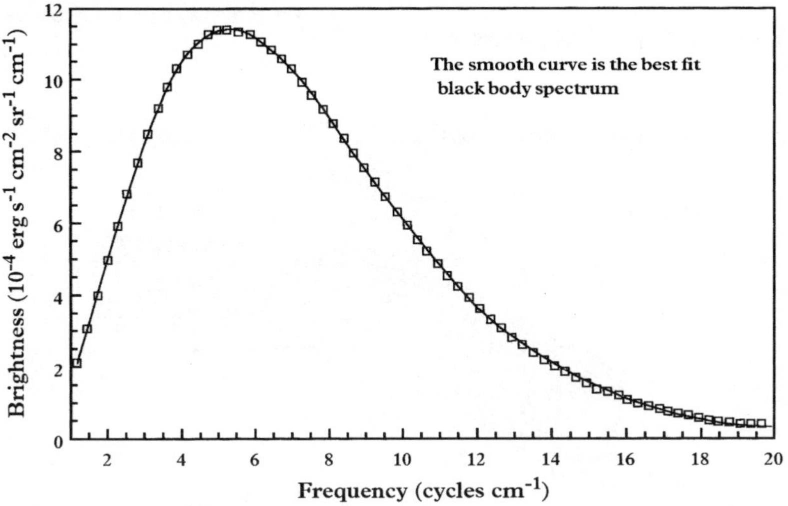

Figure 1 Initial FIRAS CMBR spectrum showing assumed 1% error bars and a blackbody curve [5]. Subsequent analysis has reduced the error bars to 0.01% [8]

spectral regions. Mechanical grating inaccuracies typically limit their use to frequencies below 100 cm^{-1} with best performance usually between 3 and 20 cm^{-1}. Then in 1969, D. Martin and E. Puplett at Queen Mary College, London, developed a new interferometer design in which a wire grid polarizer was substituted in place of the conventional beamsplitter [14]. As a result, the new design had high efficiencies over a range from 2 cm^{-1} to above 100 cm^{-1}.

Among the pioneering applications of the Martin–Puplett 'polarizing interferometer' design for astrophysics was the work by Woody *et al.* [15], who used a balloon-borne version to measure the spectrum of the CMBR from 4 to 17 cm^{-1}. Their instrument included many of the essential design elements that would later be used in the FIRAS instrument, such as a conical antenna with an apodizing horn at the input, in-flight calibration of the instrument with a moveable blackbody and a cryogenically cooled bolometer for measuring the interferogram over the instrument's full spectral range. An instrument with a similar design was flown by Woody and Richards [16, 17] with an extended spectral coverage from 1.7 to 40 cm^{-1}.

2 THE COBE/FIRAS MISSION

In November 1974, NASA issued an 'Announcement of Opportunity' inviting astronomers to propose a satellite experiment. The design of the Cosmic

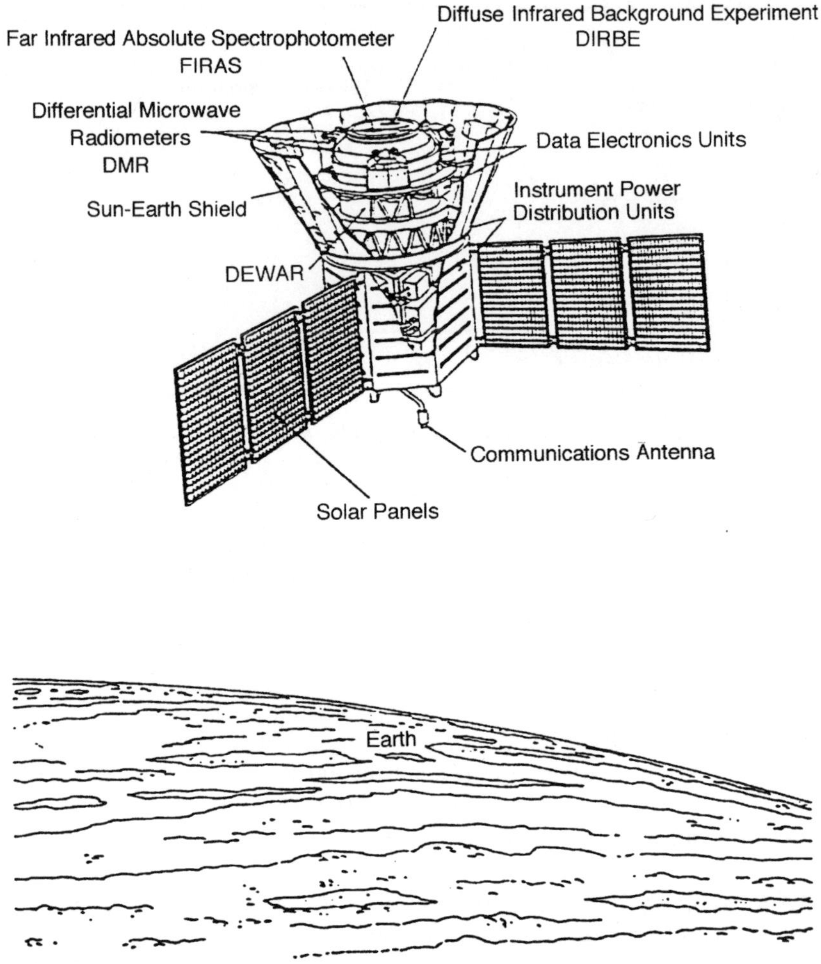

Figure 2 Artist's concept of the COBE satellite in orbit (courtesy NASA)

Background Explorer (COBE) was created in response to this Announcement by a team of seven astronomers from the NASA Goddard Space Flight Center, MIT and Princeton University. The team was led by John Mather and Rainer Weiss, and originally involved four instruments, each designed to examine a specific aspect of the cosmic background radiation: its spectral shape and temperature, its isotropy and the detection of other significant cosmic background components in the infrared. In 1976, NASA selected a team augmented with members from competing proposal groups from NASA's Jet Propulsion Laboratory and the Lawrence Berkeley Laboratory. NASA's

Goddard Space Flight Center was selected to build and launch the COBE satellite [18] and to analyze the results (Figure 2).

In addition to the FIRAS instrument [John C. Mather, Principal Investigator (PI)], COBE hosted two other experiments designed to investigate specific features of the cosmic background radiation. The differential microwave radiometer (George F. Smoot, PI; DMR) mapped the intensity of the CMBR to an ultimate accuracy of a part in 100 000 to search for anisotropy left over from the Big Bang (e.g. Ref. 19). The diffuse infrared background experiment (Michael G. Hauser, PI; DIRBE) measured the absolute infrared emission from the sky from 1 to 240 μm to search for possible cosmic infrared backgrounds due to a variety of suggested mechanisms soon after the Big Bang. The analysis of these data has proved a daunting challenge. As yet, only upper limits are available (e.g. Ref. 20) for any cosmological component to the infrared background.

Following a 15 year design and development phase, and several modifications due to launch vehicle changes, NASA launched the *COBE* on 18 November 1989 from the Vandenberg Air Force Base (AFB) into a 900 km circular orbit. At this altitude, the orbit's inclination of 99° produced a tidally induced precession of the orbital plane of one degree each day, making it Sun-synchronous. The instrument dewar for FIRAS and DIRBE is a 650 l superfluid helium cryostat developed by Ball Aerospace Division, which maintained the detectors at a temperature of 1.4 K. An Earth–Sun and radiofrequency (RF) shield protected the 2109 kg COBE satellite from direct sunlight entering the instrument portals, and provided a heat shield to minimize the thermal load on the cryostat's external shell. The FIRAS bolometers can operate effectively only below 2 K, so the FIRAS was switched off when the liquid helium supply was depleted on 21 September 1990 after 10 months of nearly flawless operation. By this time, FIRAS had surveyed over 99% of the sky, providing over 2.3 million interferograms.

3 FIRAS OPTICAL DESIGN

The CMBR has its dominant emission between 2 and 12 cm^{-1}. For efficiency in this band, the optics in the FIRAS instrument are reflective surfaces and wire gratings. The design symmetry and separated input and output ports make FIRAS a convenient differential instrument. The first input port is usually open to the sky via a 12.8 cm diameter non-imaging concentrator (Winston cone) with a 7° field of view (the Sky Horn), while the second input is connected to an internal calibration source (ICAL) whose emission is within 5% of a blackbody via a similar concentrator (the Reference Horn). The two output ports of the interferometer are identified as the left and right-side outputs.

A ray, either from the sky or from the XCAL, must first pass through the Sky Horn, which establishes multiple images of the sky at the throat of the cone. A ray originating in the ICAL is concentrated by the Reference Horn. The measured Sky Horn and Reference Horn emissivities are approximately equal, so that their contributions to the output interferograms cancel when their temperatures are equal.

3.1 CALIBRATORS AND HORNS

A significant advance over previous instruments that were used to measure the CMBR spectrum is the ability of the FIRAS to replace the sky input with a high precision external blackbody, XCAL, with a calculated effective emissivity >0.99995, based on measured material properties and explicit optical design. Based on a design by Hemmati, Mather and Eichhorn [21], the calibrator is a folded cone shaped like a trumpet mute, and has a front surface made of iron-loaded epoxy, Eccosorb CR-110, with a 25° groove angle. With this geometry, any microwave radiation received by the spectrometer from the calibrator undergoes at least seven specular reflections from the Eccosorb. This reduces the effective reflectance from the ~11% of a single normal specular reflection to $\sim(0.11)^7 = 2 \times 10^{-7}$.

Several measurements were made to test how closely the XCAL approximates an ideal blackbody. The reflectance of the XCAL was measured by D.T. Wilkinson at several frequencies near 1 and 3 cm^{-1} and the largest reflections were found to be less than 3×10^{-5}. When the XCAL is in the Sky Horn, it does not quite touch it. There is a 0.6 mm gap between the edge of the XCAL and the horn which allows some leakage on to the detectors, but this is sealed by two rows of aluminized Kapton leaves. An upper limit for this leakage was established during ground tests with a warm cryostat dome, by comparing signals with the XCAL in and out of the Sky Horn. Leakage was found to be less than 1.5×10^{-4} in the range $5 < \nu < 20\ \mathrm{cm}^{-1}$ and 6.0×10^{-5} in the range $25 < \nu < 50\ \mathrm{cm}^{-1}$. The two rows of Kapton leaves sealing the gap between the Sky Horn and the XCAL were also tested by gradually withdrawing the XCAL from the horn. No effect could be seen in flight until the XCAL had moved 1.2 cm. In short, the XCAL is an excellent approximation to an ideal blackbody that completely fills the field of view at the frequencies of interest.

The internal reference calibrator, ICAL, with a measured emissivity of ~0.96, is smaller than the XCAL. It is a cone pointed toward the detectors, so radiation from the detectors that is reflected from the ICAL will also reflect from the horn antenna before returning to the reference body for another chance to be absorbed. The calibration of FIRAS depended on both the calibrators being at known constant, uniform temperatures.

The Sky Horn has a smoothly flared aperture to reduce diffractive sidelobes over a wide frequency range. The area of the throat (0.48 cm^2) and the effective beam pattern at the throat (π sr) establish the étendue of the instrument as $A\Omega = \pi^2 r_{\text{throat}}^2 = 1.5\ \text{cm}^2\ \text{sr}$. The effective beam size on the sky can be found by equating this to $\pi^2 r_{\text{mouth}}^2 \sin^2 \theta_{\max}$, where $\theta_{\max} = 3.5°$ is the ideal half beamwidth on the sky. The étendue also establishes the long-wavelength diffraction cut-off of the instrument, since the number of geometric modes of each polarization of the radiation field accepted is $A\Omega/\lambda^2$.

At long wavelengths, the beam shape is governed by diffraction of a few modes, while at short wavelengths geometrical optics predict a flat 'top hat' angular response. In orbit, the instrument beam profile was determined by using transits of the Moon across the instrument field-of-view. The angular size of the Moon is 0.5°, which makes it an acceptable point source for beam mapping.

Lunar transits of the instrument field-of-view also indicated that the portion of the beam within the central <3.5° is not flat topped, and shows beam response variations of as much as 30%. This is only important when making measurements of sky features with angular variations smaller than 3.5°. Even then, the effect is washed out by spacecraft rotation and orbital motion during a single FIRAS integration. It is also reduced when multiple measurements are combined in the averaging operation.

Ideally, the two input horns should have zero emission and be symmetric. In fact the emission is $\sim 0.012 + 0.0015 \times (\nu/1\ \text{cm}^{-1})$ of the emission from the XCAL and is symmetric at low frequencies. The horns do not have the same pattern of emission as a function of position and angle within the beam as do the blackbodies, and so do not necessarily produce identical optical responses.

3.2 OPTICAL PATHWAYS

Instruments designed for the space environment must operate with high efficiency, use minimum power and occupy a small mass and volume. The limited volume within the COBE dewar was approximately $10^5\ \text{cm}^3$. This constraint required that the FIRAS optical path had to be folded several times as shown in Figure 3. To keep the beam from expanding beyond the size of the optics, it also had to be refocused five times.

The ray next encounters a wire grid polarizer, labeled A for (input/output) analyzer. The wires are gold-coated tungsten, stretched to half their yield stress as they are wound, and then epoxied to an Invar frame. The wires are spaced 53 μm apart and are 20 μm in diameter. Ideally, the polarizer separates two perpendicular polarizations, one transmitted and one reflected. The use of polarizing beamsplitters in the Martin–Puplett design provides good beamsplitter efficiency over a wide frequency range. The reflected sky signal on the left side is added to the transmitted ICAL signal, with the perpendicular polarization, and similarly the transmitted sky signal is combined with the reflected ICAL signal on the right side of the instrument.

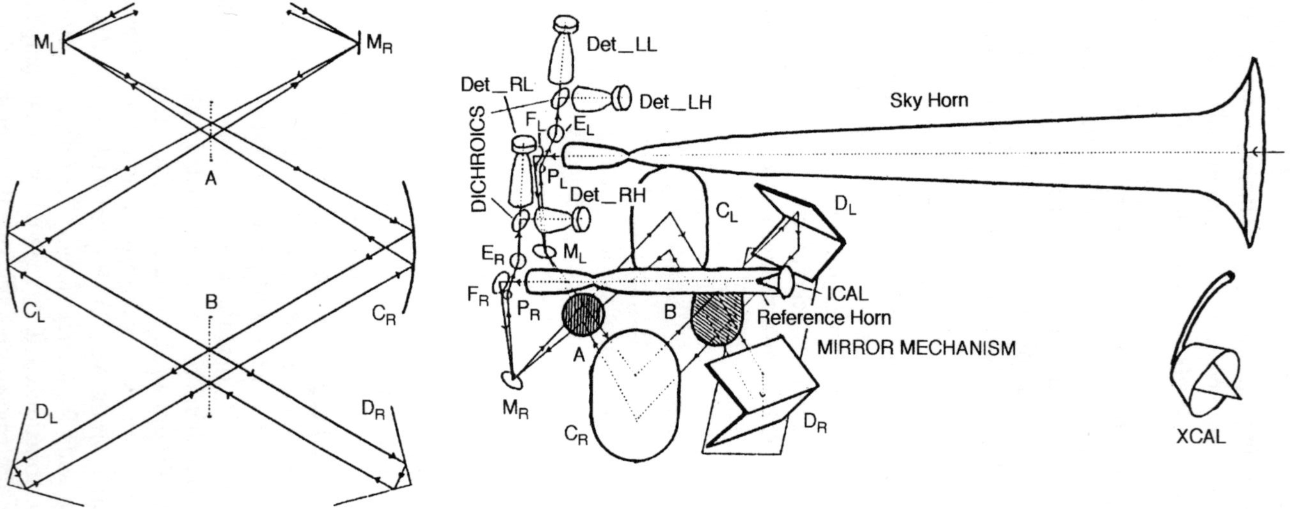

Figure 3 FIRAS optical path, folded and unfolded views

The polarized ray now encounters another off-axis parabolic collimator mirror, C_L or C_R. Each mirror has a focal point near the center of the corresponding dihedral mirrors, D_L and D_R at a distance of 40 cm, making an image of the sky at infinity, and an image of the pupil on the dihedral mirrors 40 cm farther ahead. Before reaching the dihedral mirrors, the beam encounters a second polarizer, labeled B for beamsplitter, whose wires are oriented 45° from the wires on the previous polarizer, as seen by the beam. The input beam, therefore, splits coherently into two new beams which go to the dihedral mirrors.

The dihedral mirrors shift the beam position and change the polarization, while reflecting each beam back to the beamsplitter. The polarization rotation means that the fraction which had been reflected is now transmitted, and *vice versa*. This allows the beams to recombine coherently, but with an elliptical polarization state that now depends on x, the difference in the path length from the beamsplitter to the dihedrals. After recombination, the beams retrace their path, but are now displaced in position because of the dihedral mirror shift. They are reflected from the collimator mirrors to the analyzer polarizer, which analyzes the polarization state and makes two output beams. The net result of all the combinations of polarizations, for a monochromatic source of frequency ν measured in wavenumbers, is that the ideal interferometer has the transmission functions $(1 \pm \cos 2\pi\nu x)/2$, with the sign depending on whether the output beam is on the same side or opposite the source.

Finally, after the rays leave the analyzer polarizer, they are reflected off the optical flat $F_{L,R}$ and separated from the input rays by a small pickoff mirror $P_{L,R}$. They are then focused by an elliptical mirror $E_{L,R}$, on their way to a set of dichroic filters and the apertures of the detector concentrator cones. Ray traces predict that approximately 20% of the light entering the interferometer is lost because of aberrations at the off-axis mirrors, and that there is not much improvement possible from the use of other shapes.

At high frequencies, the polarizer grids are not ideal. Their efficiencies drop, and the symmetry between inputs and outputs is no longer guaranteed by design. In addition, there is a complex interaction between the two polarizers as they form a fixed-tuned, two-beam interferometer themselves. The result is that the interferometer efficiency drops to essentially zero at certain frequencies. The input/output polarizer was therefore tilted away from the ideal position to recover the nulled frequencies.

3.3 MIRROR MECHANISM

Moving the dihedral mirrors to scan (or stroke) through a range of path differences, x, produces an interferogram, $I(x)$, which is the cosine transform of the input spectrum. Sample interferograms are shown in Figure 4 for two different temperatures for the ICAL near the sky temperature of 2.74 K. The

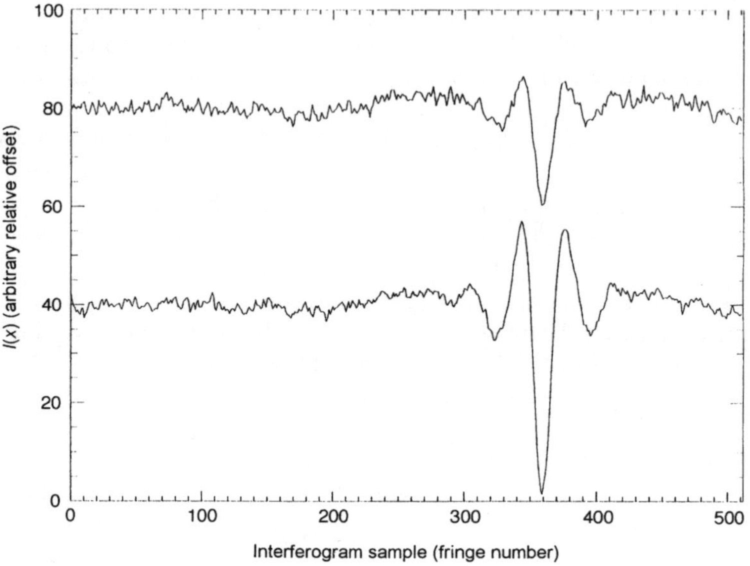

Figure 4 Two interferograms obtained at slightly different ICAL temperatures (top, 2.7514 K; bottom, 2.7639 K) relative to the sky temperature ($T = 2.74$ K)

top trace is with an ICAL temperature of 2.7514 K; the bottom uses 2.7639 K. With the ICAL set exactly to the sky temperature, the interferograms would yield a null signal in all channels if the interferometer were perfectly symmetrical and the ICAL were a perfect blackbody.

The moving dihedral mirror transport mechanism (MTM) is responsible for varying the path difference in the interferometer, and was the most difficult part of the instrument to design, build and test. In space applications, the riskiest components are those that move, and in the case of FIRAS the MTM must move almost continuously, and with high repeatability, for up to 10 months and over 10 million stroke cycles.

Although the wavelengths involved are long compared with visible light, and tolerances are increased accordingly, the cryogenic operation, mass and volume constraints made the design and construction of the MTM a challenge. A parallelogram linkage with leaf springs at the corners was chosen because it requires no lubrication, provides a smooth motion and has little friction. Standard design half inch Bendix flex pivots containing perpendicular sets of leaves in cylindrical mounts are used, but the high vibration levels of launch required the use of A718 alloy springs instead of the usual steel. To achieve the maximum path difference for a given mechanical stroke, both of the dihedral mirrors move on a single platform, in a direction perpendicular to the beamsplitter plane and 30° from the optical axes. This means that as the mirrors move, there is a lateral movement of the mirrors across the beams which has no effect on the path difference.

The scanning mechanism is driven by a specially designed linear motor with fixed coils attached to the baseplate and a moving permanent magnet to minimize the heat dissipated in the moving portion. Unfortunately, the flex pivots have very low thermal conductance. The motion is sensed by two analog methods as well as the glass scale: a magnet moving in a coil measures the velocity, and a linear variable differential transformer measures the position.

The motor is driven by a current source controlled by a linear velocity servo amplifier so that its response is independent of the motor coil resistance, which diminishes by two orders of magnitude when cooled. Two forward stroke velocities, labeled 'fast' and 'slow,' are provided, as a way of optimizing the sensitivity and recognizing potential coherent interference effects from the spacecraft. These scan velocities give path difference rates of 0.783 and 1.175 $cm\,s^{-1}$, in the ratio of 2 to 3. The commanded velocity signal is low pass filtered to reduce the forces applied during the velocity reversals and minimize excitation of spacecraft and cryostat vibration modes. Nonetheless, there is measurable evidence for these effects. The mechanical resonant frequency of the desired parallelogram mode of the mirror system is approximately 0.5 Hz, and the next important mode is at 57.57 Hz with $Q \approx 500$. This appears to be a torsional mode, excited because the force vector of the motor does not pass through the center of mass. It is the primary limitation on the servo bandwidth, so a notch filter tuned to this frequency was inserted in the feedback loop. There is also an 8.55 Hz vibration, the exact frequency depending on the mirror speed, which is probably the combination of a low-Q mechanical resonance of the cryostat and the limited band width of the mirror mechanism servo loop.

The position of the mirrors in their stroke is measured by counting fringes on a reference ruler on a glass scale with a fringe spacing of 20 μm at room temperature. The scale is illuminated by an optical fiber from an external light emitting diode, and a segment of a spherical mirror creates an image of the scale back on itself to create an optical lever. Light is transmitted through the scale twice and is returned to a fringe detector outside the cryostat by another fiber optic line. Because of the angle between the optical axes and the direction of stroke, the change in the optical path difference is 40 cos 30° μm per fringe. The accuracy of translation of this mechanism was measured with a laser interferometer, and the orientations of the flex pivots were adjusted slightly to minimize tilts and rotations during the stroke. The relevant parameters of the MTM scanning system are given in Table 1.

In the forward direction, a counter is set by the signal from the reference marker and then fringes are counted until the desired stroke length has been achieved and the logic reverses the stroke direction. Two stroke lengths are provided: a 'short' stroke of 512 fringes and a 'long' stroke of 2048 fringes. The long stroke has a potential fourfold improvement in the frequency

Table 1 Parameters for a single scan

Mode	Scan time (s)	Flyback time (s)	Total time (s)	A/D Samples per scan
Long–fast	6.2	1.4	8.1	4104
Long–slow	9.2	1.4	11.4	6156
Short–fast	1.5	0.5	2.5	1026
Short–slow	2.3	0.5	3.4	1539

resolution of the instrument. The zero path difference location is at the 355th fringe, thus its location is not symmetric in either scan length. In the reverse direction, the scan speed is $1.46\ \mathrm{cm\,s^{-1}}$ ($= 5\ \mathrm{cm\,s^{-1}}$ path difference). The reverse motion continues until the reference mark has been passed, at which point the commanded velocity returns to its forward value.

The fringes on the reference scale are the source of the pulses used to sample the interferograms at the correct locations. More frequent pulses are needed, so the marks from the scale are subdivided in time by 2 if the mirror velocity mode is fast, or 3 if it is slow, using a phase-locked loop with a frequency divider (counter) in its feedback circuit. The bandwidth of this loop is not wide enough to respond to the 57 Hz MTM vibration mode or to track completely the 8 Hz MTM vibrations excited by the mechanism. The timing of the sample pulses gives information about the vibrations and enables the accuracy of the sampling to be estimated.

The FIRAS instrument operated in an environment in which cosmic ray hits were frequent. At times, these events affected the MTM zero reference mark photodetector so that the MTM was driven past its electronic stop and into a physical stop. This did not damage the MTM, but the linear motor driving it dissipated a power of ~200 mW for a few tens of seconds until a watchdog timer circuit restarted the scans. This heat then diffused into other parts of the instrument, at times heating them to temperatures as high as 5 K. Most occurrences were in high-radiation regions known as the South Atlantic Anomaly (SAA). After December 1989 the MTM was held in position during the SAA passages, but this still increased the heat dissipation of the MTM motor, so after March 1990 the MTM was turned off during those passages. Because of these 'end of travel' events in the early part of the flight, the temperatures of the mirrors and wire grids cannot be treated as constant. Thus radiation from the MTM and mirrors was ultimately included in the calibration model. The data obtained in the 'long–slow' mode were particularly affected by the end of travel events, and observations made with this mode were not used.

3.4 OPTICAL FILTERS AND SIGNAL DETECTION

Each of the two output beams is split by a dichroic filter into a reflected high-frequency (20–95 cm^{-1}) beam and a transmitted low-frequency (1–20 cm^{-1}) beam, for a total of four output beams designated 'right–high', 'right–low', 'left–high' and 'left–low'. The intensity of each beam is measured by a bolometer detector which is sampled from 1024 to 6144 times per stroke, depending on the length and speed of the MTM stroke. There are two scan lengths, two speeds and four detectors, yielding 16 possible output formats such as 'short–slow : left–high' and 'long–fast : right–low'.

The dichroic filters are multilayer capacitive grid filters with aluminum squares, photoetched on polyethylene sheets which are transparent at the FIRAS operating wavelengths. The nominal separation frequency is 20 cm^{-1}, but the high-frequency detector also receives some low-frequency radiation. In addition, each detector has a low pass capacitive grid filter across the large aperture of its concentrator cone. In the case of the low frequency detectors, the low-pass filters are identical with the dichroic filters, but for the high frequency detectors the cut-off frequency is 100 cm^{-1}. The concentrators are elliptical concentrators matched to those on the input antennas, and have the same throat size and étendue.

The four detectors have been described previously by Serlemitsos [22] and are composite antimony-doped silicon bolometers with electrical noise equivalent powers (NEPs) of $\sim 4 \times 10^{-15}$ W $Hz^{1/2}$. The operating temperature of the bolometers is 1.58 K during normal observations, but rises as high as 1.73 K during calibrations with high temperature sources. The FIRAS required bolometers with short time constants (3–40 ms), high throughput (1.35 sr cm^2), and low NEP. They also had to be spaceflight qualifiable, with very small susceptibilities to microphonic interference. The Goddard Space Flight Center developed these bolometers since none that met the stringent flight requirements were available through outside suppliers.

Each bolometer is a diamond octagon 0.78 cm across the corners and 25 μm thick. The size is chosen to match the étendue of the antennas and cones, and the diamond is blackened with a chrome–gold film. The optimum film surface resistance for FIRAS bolometers in their chambers was calculated to be 267 Ω per square, which was achieved with 20 Å chromium, 5 Å chromium–gold and 30 Å of gold. Each bolometer has a silicon thermometer glued to a corner with Epo-Tek H20E silver epoxy, and is thermally anchored with very thin brass lead wires. The bolometers are supported by crossed pairs of taut Kevlar fibers, 6 μm thick and tensioned to 0.25 N, from an Invar frame (since Kevlar expands on cooling). The fibers are epoxied into place with Stycast 2850 FT, which was done before vacuum deposition of the absorbing film.

The type and size of the thermal leads depends on the desired time constant of the bolometer. The low-frequency 'slow' bolometers are made with a smaller thermal conductance and longer time constant (40 ms) using 6 μm PtW wire. For the high-frequency 'fast' bolometers (6 ms), 8 μm brass wire was used. The thermal conductivity of the Kevlar fiber is 20 T^2 μm W cm^{-1} K^{-1} for a total conductivity of 0.4 nW K^{-1}. The total conductivity of the slowest bolometer was greater than 50 nW K^{-1}, so the fibers contributed less than 1%.

The two sides of the interferometer have nearly identical detectors and filters, but the optical responses of the two sides are dramatically different as a consequence of either the degradation of the detector absorbing film or defective filters. The latter possibility seems unlikely because the spectral responses were measured to be similar. On one side, the low-frequency detector is good and the high frequency detector is poor, while the opposite is true on the other side. The ratio between the good and poor detector responses is about 5.

The detector voltage is measured through a JFET source follower amplifier (gain ≈ 0.995) mounted near the detector but thermally isolated from it. The amplifier uses dual JFETs connected in parallel, and suspended by Kevlar threads inside a copper can, so that the JFET bias power can heat it to an operating temperature of ~70 K. Wires into the bolometer housing are filtered against radio and microwave energy by being cast into an Eccosorb CR-124 iron-loaded epoxy disk. Nevertheless, the detectors or possibly the preamplifiers were sensitive to radio frequency emitted by the microprocessor clocks (~4 MHz). Beat frequency oscillations between different clock signals were also seen as excess noise levels at those frequencies despite efforts to provide additional external decoupling.

The space environment is hostile to digital circuitry, and especially to high quantum efficiency infrared detectors. Increasing the detector area to improve its light-gathering ability also increases its cross section to impacts by cosmic rays and other particle fluxes produced by the Sun and celestial sources. The natural Galactic cosmic ray rate is ~4 cm^{-2} s^{-1} depending on the effectiveness of the Earth's magnetic field in shielding them. Most of these are at very high energy (GeV and higher), which results in over 50 hits per interferogram. In addition, trapped particles in the Earth's field are much more abundant than the Galactic flux of cosmic rays in certain portions of the orbit. The radiation belts have a distribution that dips down to low altitudes near the COBE orbit, in the South Atlantic Anomaly. The flux of energetic protons in this region is so intense that no useful FIRAS data can be obtained. In the remainder of the orbit, particles produce abrupt signals in the bolometers which decay with the characteristic detector time constant. Individual bright hits, dubbed glitches, can be modeled and removed by software. The lower amplitude (and more numerous) glitches are not individually detectable, but are responsible for ~80% of the total in-orbit 'detector' noise. A typical interferogram has 10 recognizable glitches removed, which is about 2% of the total number of points in each interferogram. Even after 'deglitching' is performed, the in-orbit rms noise is twice as high as the noise observed during ground-based tests. Details on the deglitching algorithm are given in Section 5.1. Each bolometer is mounted in a housing with a short aluminum tube between the concentrator throat and the detector, to reduce the view factor for cosmic rays and to attenuate RF fields from the transmitter. Machined tantalum shields capable of stopping protons of 50–100 MeV (depending on direction) surround the rest of the detector as an additional source of cosmic ray attenuation.

3.5 MECHANICAL DESIGN AND COBE SPACECRAFT INTERFACE

Except for the Sky Horn and the XCAL, the optical components, moving mirrors and detectors are all built into a box roughly 50 cm on a side, constructed on a baseplate which is in turn attached to the dewar interface flange. This box is surrounded by a cylindrical instrument 'Interface', which supports the Sky Horn and XCAL, and is joined to the baseplate at a flange. The lower portion of the Interface is closed off with a cover to protect the interferometer from stray infrared inputs. This cover has two holes, one for each horn antenna, with labyrinth seals of aluminum and Eccosorb CR-110 infrared absorber to allow air to escape while blocking infrared radiation.

The entire mechanical structure and all the mirrors and cones are built from 6061-T6 aluminum alloy, so that upon cooling to liquid helium temperatures no differential contractions are introduced that could affect the beam geometry. The cooling process itself produces thermal gradients that could produce strain or slip at bolted joints. To avoid this, the cooling rate was limited so that the cool down to 4 K took 2 days. Bolt threads were coated with Solithane to prevent loosening under vibration. Belleville washers were used throughout to take up the differential contraction of the steel bolts and the aluminum structures. Where needed, thermal insulators of G-10 epoxy fiberglass were used, and relative motions of the parts were minimized by compensating lengths of fiberglass elsewhere.

In the case of the polarizing beamsplitters, it is not possible to use aluminum frames, because aluminum would contract more than the tungsten wires and relax their tension. Therefore, the polarizer frames are Invar, and a kinematic mount is provided with spring-loaded slip joints and adjusting screws. This arrangement caused alignment problems, since the vibration levels during test were sufficient to cause erosion of the kinematic mounts. However, the alignment was well preserved even under the stresses of launch, on the basis of a comparison of the high frequency instrument responsivity before and after launch.

The Sky Horn antenna is supported primarily at the top end where it is mounted on a platform inside the Interface with a G-10 thermal insulator. Its lower end is mounted in a loose slip joint near the input to the spectrometer with minimal contact area and pressure to reduce heat flow along the horn. This was successful in that the thermal gradient along the horn was negligible. The XCAL mechanism is attached to the same platform on the Interface, and includes a stepper motor with redundant coils, an 8 : 1 gearbox and ball bearings.

Stepper motor controllers are provided for three motors: to move the XCAL, to latch the calibrator and to latch the mirror mechanism. For each motor there is a prime and a redundant controller with its own windings on the motor, and a command to slave the redundant to the prime if both are powered on simultaneously.

4 TEMPERATURE CONTROL AND THERMOMETRY

All of the principal input sources for the FIRAS (the XCAL, ICAL and horn antennas) are temperature controlled. Redundant temperature controllers were used for each item, consisting of a resistive heater, a germanium resistance thermometer (GRT) and associated electronics. There were 22 separate GRTs, for the controllable items and various locations throughout the rest of the instrument, including pairs on each of the four detector housings, the dihedral mirrors and other elements of the instrument structure. The GRTs were calibrated to 1 mK accuracy against a transfer standard GRT from the National Institute of Standards and Technology, but measurement of other devices from the same batch showed calibration drifts up to a few mK after several years.

The performance of the controller was strongly dependent on the heat capacity and internal thermal conductivity of the object to be controlled. At the highest temperatures used during the calibrations (>15 K), the calibrator's equilibration times were tens of minutes, and the time required to cool down to the lower temperatures used for sky observations was as long as 1 h. At the lower temperatures, the control loop time constants and the thermal time constants were very short, of the order a few seconds, and in general the temperature stability, when not subject to abrupt environmental perturbations, was better than 0.2 mK.

All the thermometers were used in a four-lead configuration to eliminate sensitivity to the cryostat lead resistances which were ~64 Ω each way. The thermometers were excited by stabilized sine wave current sources operating near 40 Hz with a choice of currents. The voltage developed across each GRT was amplified by a wide band differential amplifier and synchronously demodulated before DC amplification. The analog to digital converter provided 14 bits of resolution, equivalent to 0.05 mK per bit at 2.7 K. Each GRT was read out by one of two digitizers through MOSFET multiplexers, which cycled through the GRTs and four standard resistors to provide frequent calibration. The stability of the preamplifier and the ADCs, determined by comparison with the calibration resistors, was good enough that no significant trends were seen over the course of each day. In addition, there was no evidence of thermometer self-heating at these currents and accuracy levels.

Temperature gradients within the XCAL and ICAL calibrators and horns are minimized by heating each one only at its single point of mechanical support. The ICAL temperature is measured by two GRTs and controlled by a third. The XCAL temperature is measured by three GRTs and controlled by a fourth. The scatter among the readings of these GRTs is 9 mK for the internal calibrator and 3 mK for the external calibrator, when these bodies are set at 2.7 K. Most of this observed variation is probably due to post-assembly changes in the GRT calibrations.

Adjusting the ICAL temperature to balance the sky input from the CMBR reduced the net signal by 99%. This differential approach is responsible for the ability of FIRAS to detect deviations of the CMBR from a perfect blackbody less than 0.005% (rms) of the peak. Absolute measurements of the sky flux require knowing the emission of the ICAL and other components of the instrument. This is done by examining the changes in the observed spectrum with variations in the component temperatures.

For the ICAL and two horns, variation is produced by commands to attached heaters. The overall throughput of the instrument is measured by placing the XCAL in the Sky Horn aperture, which thereby replaces the sky signal with a known input. The XCAL signal is known because it is a blackbody to within 0.003% according to calculation and ground measurement. Errors in the thermometry of the XCAL are the limiting factor when comparing the absolute sky flux measurements of FIRAS with other instruments, but do not affect the determination of the distortion of the CMBR spectrum.

To check for systematic variations that might affect the data acquisition and post-flight calibration process, the operational mode of FIRAS was varied throughout the mission. The MTM scan speed was switched between its slow and fast modes every other orbit to allow instrumental errors to be distinguished from sky signals. At the same time the scan length was switched between short and long to provide higher spectral resolution at the expense of signal to noise. In addition, the temperature of the internal black body reference (ICAL) was switched between a 'sky null' setting and a setting 12 mK hotter every 3–4 days to track possible variations in the instrumental gain between the calibrations.

5 DATA PROCESSING

Some processing of the data is done on the spacecraft to reduce the data transmission rate to the ground. First, AC coupling is used to reduce the scale of the bolometer signals and permit high preamplifier gains. Also, there is a pole-zero compensator (a filter with a transfer function that compensates for the bolometer time constant) to shorten the pulses resulting from cosmic ray glitches to be discussed in the next section. The resulting electronic transfer function is a five-pole Bessel low-pass filter with a −3 dB bandwidth of 100 Hz, and a six-pole high-pass filter that is −3 dB at 0.3 Hz.

The detector signals are sampled at equal intervals of path difference as indicated by the output of the mirror mechanism phase lock loop, slaved to the mirror position fringes. The sample rate is ~681.43 Hz, which is enough to avoid significant aliasing of high frequency noise into the infrared signal bandwidth (1–100 Hz). The digitized preamplifier outputs are then processed with an SBP 9989 16-bit microprocessor, one for each detector output. Under normal operations, the microprocessor applies a digital low pass filter to the incoming data, and between 2–12 samples are averaged together to produce an interferogram (IFG) 512 points long. The microprocessor then adds together a set of either four long scans or 16 short scans to form a group for telemetry to the ground. Double buffering is used because the data transmission is asynchronous with the digitization. The microprocessor also runs a pulse finder algorithm on the original, unaveraged samples to identify glitches from cosmic rays and flags all points where a glitch has been located in any one of the set of scans. While taking long stroke data for the high-frequency detector channels, the microprocessors only sample the initial quarter of the stroke, producing an interferogram almost identical with the short-stroke data, although one quarter of the duty cycle. If an entire long stroke is accumulated with a 512-point interferogram, the Nyquist frequency (36 cm^{-1}) is midway through the pass band of the high-frequency channels. Any of the details of the operation of the microprocessors were determined by values contained in Table 2; the first four are for the low frequency channels and the second four for the high-frequency channels. The number of samples needed to clear digital transients from the filters is indicated in the second row of the table. Also provided was a choice of basic program modes that enabled or disabled the on-board digital filter, glitch location software, and glitch removal software. Diagnostic modes were also provided, in which the raw, unfiltered data could be examined, the intervals between samples could be measured to a 10 μs resolution and the microprocessor memory could be checked or reprogrammed. Despite the harsh environment, no evidence was seen of microprocessor operation errors in orbit caused by cosmic rays.

Approximately 2.3 million IFGs were collected from FIRAS. Of these, 24% were unusable owing to transmission losses, saturation while crossing the Galactic plane or possible interference from the Earth limb, Sun and Moon. Engineering and diagnostic data accounted for 10% and instrument calibration data accounted for 10%, leaving 54% for actual sky data amounting to about 1.4 million IFGs. The FIRAS was in the 'short–slow' mode about 50%, the 'long–fast' about 40% and the 'short–fast' about 10% of the time.

Table 2 Microprocessor set-up table

	Low frequency				High frequency			
Mode	LF	LS	SF	SS	LF	LS	SF	SS
Sample/group	8	12	2	3	2	3	2	3
Samples to flush	8	9	2	3	9	9	2	3
Groups/sweep	512	512	512	512	512	512	512	512
Sweep/IFG	4	4	16	16	4	4	16	16

The ground-based analysis phase began with processing the flight data to extract the instantaneous pointing direction of the FIRAS beam. The FIRAS pipeline software system also checked the instrument state data, microprocessor information, engineering data, status values and attitude quantities associated with the interferograms. The science and engineering data were collected asynchronously to avoid coherent crosstalk, so the pipeline software had to correlate the time-tagged science data and engineering data from two separate streams in order to perform the quality checking.

At the first stage of the pipeline, a preprocessor performs the correlation of data streams and calculates the midpoint of the IFG collection time which is later associated with the spacecraft attitude solution to derive a boresight position, and ultimately a sky pixel address corresponding to each IFG observation. The preprocessor uses the data transmit time in the science record to extract the scan speed and stroke length from the corresponding engineering and status record.

The individual IFGs which passed a stringent quality checking procedure were then grouped for co-addition. At this stage, data records must have solar aspect angles >91.2°, earth limb aspect angles >87°, lunar aspect angles >22°, a dihedral temperature <3.5 K and an ICAL temperature within 2 mK of one of the specified temperatures. All IFGs that satisfied these criteria were assigned to the co-add group for that ICAL temperature bin. Each IFG was then divided by the real-valued commanded preamplifier gain and the number of onboard scans, since the digitized voltages are proportional to these quantities.

5.1 DEGLITCHING

As discussed in Section 3.4, charged particles incident on the FIRAS bolometers give rise to 'glitches' in the interferograms. The charged particles deposit a spike of energy in the detectors, giving rise to a delta function response on the part of the detectors. Each glitch produces an abrupt temperature rise at the detector, and the detector and preamp filters respond with a well defined shape, which for the more energetic (and thus most damaging) hits can be recognized and removed. We remove the glitches using a one-dimensional variant of the radio astronomers' CLEAN procedure [23] to subtract glitch profiles iteratively from each interferogram.

The glitch spectrum is the product of the instrument electronics transfer function and the detector time constant:

$$GS(\omega) = \frac{Z(\omega)}{1 + i\omega\tau} \tag{1}$$

where ω is the audiofrequency of the instrument, $Z(\omega)$ is the electronics transfer function and τ is the detector time constant discussed in Section 3.4. The audiofrequency ω in rad s^{-1} is related to the optical frequency ν in cm^{-1} times the MTM scan speed v in cm s^{-1}: $\omega = v\nu$. The glitch profile in the time domain is the inverse Fourier transform of the glitch spectrum:

$$GP(x) = \Omega^{-1} \int_{-\infty}^{+\infty} d\nu \, e^{-2\pi i\nu x} \frac{Z(v\nu)}{1 + iv\nu\tau} \tag{2}$$

where Ω is the normalization of the transform.

The co-addition program uses the glitch profiles in the time domain to 'deglitch' the interferograms. The glitch profiles have been normalized so that they have zero integral and peak values of unity. The glitch profiles are dependent on the MTM scan speed, on the channel and on the position of the glitch within the raw data stream. Consequently, there are 26 profiles for each channel–scan speed combination.

Before executing the CLEAN algorithm, a statistically robust estimate, called a template, of the 'correct' interferogram is derived from an ensemble of interferograms by a mid-averaging technique. This template is itself a good estimator of the interferogram and is immune to the effects of large glitches, but its noise level can be further improved by 10% with the CLEAN procedure. This template is first subtracted from each individual interferogram to obtain a residual $I_{res}(\delta)$ composed primarily of detector noise and glitches. In some cases the residual interferograms are then seen to have noticeable peaks at the zero path position, indicating real changes of the infrared signal from the mean. For this reason, a second template is created by a statistically robust estimator of $I_{res}(\delta)/I_{res}(0)$. This second template is fitted to each residual interferogram with an adjustable amplitude, and then subtracted to form a second residual interferogram if the primary and secondary templates exceeds some threshold. The thresholds are adjusted to make the deglitcher remove glitches at the same rate at the peak of the interferogram as other places in the interferogram.

A modified CLEAN algorithm is then executed on this second residual until all glitches larger than 3.5σ have been removed. In this context σ is defined as

1.25 times the median absolute value of the residual noise. Since cosmic ray glitches are observed to have an E_g^{-3} differential number distribution, where E_g is the glitch amplitude, residual events are numerous but small, and there is little to be gained by more sophisticated algorithms.

5.2 *BASELINE CURVATURE*

The stroke length is not small compared with the depth of focus of the interferometer, and ray traces show that the amount of unmodulated background power that reaches the detectors varies by ~0.5% over the stroke. This produces a very low frequency modulation of the interferogram that masks true infrared signals at those same frequencies. This modulation is proportional to the total power coming through both input ports (and not their difference). The low-frequency modulation represented by this baseline curvature produces responses at higher frequencies when the baseline is apodized and Fourier transformed, and is effectively a zero frequency spectrum ghost with sidelobes. To extend the accuracy of the instrument to the lowest possible optical frequencies, the baseline curvature was removed in a physically realistic manner. Measurements showed that the baseline curvature was well represented by a quartic polynomial, which was subtracted before Fourier transforming the incident interferogram. This also subtracted the lowest frequency points of the discrete Fourier transform spectrum.

The remaining noise IFGs in the ensemble (provided there are two or more) were averaged, and it was at this point that the baseline curvature effects were removed with a fourth-order polynomial baseline fitted to the co-added IFG using a least squares fit. Last, a transient response from the onboard digital filters was subtracted from the first 128 points of the co-added IFG using a least-squares fit.

At the same time as co-addition, each IFG was Fourier transformed and the gain factor of the internal electronics was applied to estimate its complex uncalibrated noise spectrum. The covariance matrix of the spectrum associated with the co-added IFG was calculated from these spectra. The diagonal of this covariance matrix is an estimate of the variance at each frequency point of the uncalibrated average spectrum. For each channel and scan mode, an overall combined covariance matrix for the whole mission was formed from the weighted total of the matrices from each ensemble of IFGs. The relative magnitude of the off-diagonal terms was less than 1%.

The co-added IFGs are apodized with an apodization function that takes into account the range of path differences actually sampled by FIRAS. Then they are Fourier transformed to produce complex-valued spectra. The spectra are converted from counts to volts at the detector upon division by the electronics transfer function and by the analog-to-digital converter scale factor.

6 CALIBRATION

6.1 THEORY

The input sky interferogram is modeled as a function of optical path difference by the equation

$$I(x) = \int_{-\infty}^{+\infty} \mathrm{d}\nu \, \frac{Z(\omega)}{B(\omega)} \sum_{i=1}^{N} E_i(\nu) S_i(\nu) \cos(2\pi\nu x) \tag{3}$$

where x is the optical path difference, ν is the optical frequency in cm^{-1}, ω is the audio frequency in rad s^{-1}, $Z(\omega)$ is the electronics transfer function, $B(\omega)$ is the bolometer response function, $\mathrm{E}_i(\nu)$ is the effective emissivity of the source object and $S_i(\nu)$ is the spectrum of the source object. The N source objects in the input signal are the sky and various instrument components (primarily the internal calibrator).

A coadded sky interferogram, $I(x)$, is apodized and Fourier transformed into a spectrum in terms of digital counts:

$$Y(\nu) = \int_{-\infty}^{+\infty} A(x) I(x) \, \mathrm{e}^{2\pi i \nu x} \, \mathrm{d}x = a(\nu) * \left[\frac{Z(\omega)}{B(\omega)} \sum_{i=1}^{N} E_i(\nu) S_i(\nu) \right] \tag{4}$$

where $*$ is the convolution operator, $A(x)$ is the apodization function, $a(\nu)$ is its Fourier transform and the other factors are defined as in equation 1. The spectrum is normalized by the electronics transfer function to yield a spectrum in volts.

The bolometer response function is computed for the spectrum based on its commanded bias, readout voltage and bolometer temperature. The bolometer response function, $B(\omega)$ and the optical transfer function, $H(\nu)$, are used to convert the spectrum from volts to My sr^{-1}, yielding a differential calibrated spectrum of the sky measured as a spectrum of the difference in signal between the sky and the internal calibrator:

$$D(\nu) = \mathrm{e}^{i\nu\phi} \, H(\nu)^{-1} \left[\frac{B(\omega)}{Z(\omega)} \, Y(\nu) - C(\nu) \right] \tag{5}$$

The vibration and harmonic corrections, $C(\nu)$, are then applied to the differential spectrum. The full discussion of the $C(\nu)$ correction is given by Fixsen *et al.* [24] and in the COBE Explanatory Supplement [3]. These corrections are found by fitting models to the extensive calibration data sets, and are essential to the accuracy of the calibration. There is a small phase jitter in the FIRAS data which is corrected in each differential spectrum by the application of a linear autophase correction ϕ, which is computed to minimize the imaginary part of the spectrum.

The instrument emissivities are used to compute a spectrum of the internal calibrator using Planck functions $P(T, \nu)$ of the measured temperatures of the instrument components for the source spectra:

$$R(\nu) = H(\nu)^{-1} \sum_{i=1}^{N-1} E_i(\nu) P_i(T_i, \nu) \tag{6}$$

where $N - 1$ signifies that the sky source is not included in the spectrum.

The differential sky spectrum and the internal calibrator spectra are combined to produce a calibrated spectrum of the sky:

$$S_{\text{sky}}(\nu) = e^{i\nu\phi} H(\nu)^{-1} \left[\frac{B(\omega)}{Z(\omega)} Y(\nu) - C(\nu) \right] - H(\nu)^{-1} \sum_{i=1}^{N-1} E_i(\nu) P_i(T_i, \nu) \tag{7}$$

6.2 *SPECTRAL RESOLUTION*

A fundamental limitation of the performance of the FIRAS is the divergence of the optical path. The beam divergence at the dihedral mirrors has a 14° full width that modifies the path difference relative to the principal ray by the factor $\cos \psi$, where ψ is the angle from the principal ray. For an ideal instrument, this effect can be calculated by integrating this factor over a circular beam. The variation in the path difference effectively modulates the input interferogram by an 'instrument apodization function,' which is equivalent to an effective spectral window function of full width $\Delta\nu/\nu = 1 - \cos \psi_{\max} = 0.007$. The mean apparent frequency is also shifted downward by half this amount. This ideal resolution is complicated by the fact that, as measured by the beam profile analysis of the lunar transit data, the true instrument angular response is not uniform. The general effect can be directly observed in the interferogram for monochromatic sources.

Fortunately, nature provides two narrow emission lines from ionized carbon [C II] and nitrogen [N II] in the interstellar medium that are bright enough to provide a direct in-flight measure of the FIRAS spectral resolution at two frequencies, 63.395 and 48.72 cm^{-1}. The measured instrument full width at half maximum resolution derived from these lines ($\Delta\nu/\nu \approx 0.0035$) was found to be substantially better than the expected value above.

6.3 *APODIZATION*

The finite range of the path differences is represented by an apodization function, $A(x)$, which multiplies the interferogram and must be zero outside the measured range [0, 512]. The interferogram peak has been chosen so that it is not centered in the measured range, in order to obtain greater spectral resolution. The inverse Fourier transform of the apodization function gives the

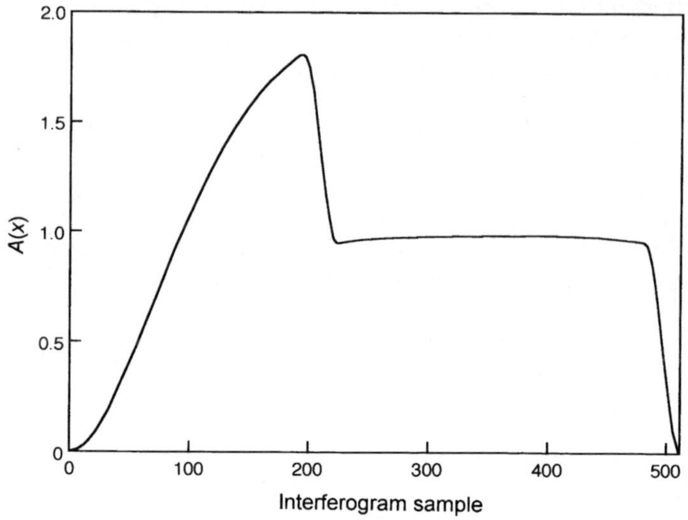

Figure 5 Example of short-stroke apodization function, *A*(*x*). The zero path difference point is at 355 counts

spectral resolution window function. The apodization function used for the short stroke data is shown in Figure 5.

The selected function is the product of two factors, $A_1(x)A_2(x)$. The first is symmetrical about $x = 0$ and declines smoothly to zero at the long end of the interferogram. The second factor declines smoothly to zero at the short end of the interferogram, has $\mathrm{d}A_2(x)/\mathrm{d}x = 0$ at $x = 0$ and has $A_2(x) + A_2(-x) = 2$. This second factor permits the use of both sides of the interferogram peak for broad-band photometry and phase determination, while eliminating discontinuities associated with the truncation of the interferogram at the short end.

The Fourier transform of the apodization function gives the instrumental spectral window function before the calibration gain correction is applied and is the dominant factor in determining the instrument line profile. The full width at half maximum of the Fourier transform of $A(x)$ is 0.7046 cm^{-1} (21.12 GHz) for the low resolution spectra and 0.2145 cm^{-1} (6.431 GHz) for the high resolution spectra.

6.4 ZERO PADDING

The Nyquist frequency set by the sampling rate is 145.212 cm^{-1}. The use of a 320-point spectrum leaves 43 significant elements in the 2–21 cm^{-1} frequency range. The excursion of the mirrors is not symmetric about the zero point, with about 1.2 cm in one direction and 0.5 cm in the other in short strokes. Hence there is information down to 0.4 cm^{-1} resolution rather than only to 0.56 cm^{-1}. A 320-point spectrum may be obtained by extending (padding) the short side of the IFG to make a total of 640 points. This number was chosen to make a well behaved covariance matrix for the real part of the spectrum. In principle, the imaginary part of the spectrum is zero if the interferogram is symmetrical. The FIRAS interferograms are not symmetrical because the apodization function is not symmetrical. The real part of the spectrum already contains almost all the spectral resolution to which we are entitled by the assumption of symmetry.

6.5 CALIBRATION DATA SETS

The important calibration data were taken in 16 calibration periods. In the first part of the mission, the calibrator was commanded into the horn approximately 2 days each month. However, for the last 6 weeks of the mission the calibrator was commanded into the horn approximately 3 days each week. Thus approximately 10% of the data set are calibration data. Data were collected with the XCAL in the horn and at a temperature of ~2.2 K, which is close to the minimum temperature achievable by the controller. Then, the XCAL was raised to 2.7, 3.5, 6, 8, 10 and 15 K. Interspersed with these measurements, the ICAL was raised to 6 K and then, in steps, to 15 K. Also the Sky and Reference Horn temperatures were raised to 6 K. During the last 40 days of the 10 month FIRAS mission, it was realized that the lack of calibration data would be the limit on the ultimate sensitivity, so seven additional periods of calibration data were added with the XCAL, the ICAL and the horns near 2.7 K. Although the four detectors, two scan speeds and two stroke lengths yield 16 potential data sets, the 'long–slow' mode was not used, so that only 12 unique data sets resulted.

A total of 1 411 829 good (free from transmission errors) 512-point sky IFGs were received. Approximately 20% of the sky data are rejected for various reasons (e.g. Sky Horn temperature, high noise, bolometer temperature too high). After sorting into the 12 data sets, 11 mission periods, 6 144 pixels and several instrument states, the remaining 1 138 149 IFGs are sorted into 272 525 groups with an average of about four IFGs per group or co-add. There are an average of 187 IFGs per pixel corresponding to an integration time of ~30 min.

There are 149 260 calibration IFGs. Approximately 10% of the calibration IFGs are rejected. The fraction is lower for the calibration data because they were primarily collected near the end of the mission when more experience had been obtained in operating FIRAS. The remaining 134 739 IFGs are sorted into 3028 co-adds with an average of 44 IFGs per co-add. The calibration co-adds are larger because they are not divided into pixels. The maximum size co-add is arbitrarily set at 100.

In the initial part of the ground-based calibration processing, calibration data and sky data are treated essentially identically. The co-added IFGs are apodized, then Fourier transformed, to produce complex-valued spectra. The spectra are converted from counts to volts at the detector upon division by the electronics transfer function and by the analog-to-digital converter scale factor. At this point, the processing of the calibration data and the sky data diverges. The co-added calibration IFGs (and their associated covariance matrices) are now used as input to the calibration model program to derive a solution, whereas both the co-added sky and calibration IFGs are calibrated with the resulting solution.

6.6 MODEL FITTING

The calibration procedure outlined by Fixsen *et al.* [24] includes ~3000 parameters for each detector and scan mode. Some of the parameters are non-linear and the process involves considerable computation. Rather than weighting according to the variance of each spectrum at each frequency, the mission average noise spectrum was used for the noise spectrum shape (dependence on frequency) for each type of spectrum (particular setting of temperatures) and the glitch weighting was used to scale the noise of each co-add.

The truncated long and short IFGs are calibrated together. The data taken with the external calibrator in the input aperture were used to construct a calibration model for each channel. Each model includes seven parameters to describe the bolometer, two parameters for the JFET amplifier, six parameters for the spectrometer and five parameters for the unwanted vibrations. The temperature of each of the four 'controllables' and an overall time shift for each co-add are also parameters of the fit, with constraints consistent with the thermometer readout noise (temperature dependent). A complex number at each frequency parameterizes the emission from each of seven sources: internal and external calibrators, sky and reference concentrators, mirrors, structure and bolometer housing, with the Kirchhoff constraint that they sum to zero. The complex number for the external calibrator is identified as the transfer function, as the external calibrator is black and fills the beam. The fitted calibration models are then applied to both the sky and calibration spectra, resulting in sky co-adds and calibration co-adds in absolute flux units.

The fitting procedure uncovered a systematic trend in the fitted ICAL temperature. The magnitude of this trend is ~6 mK with most of it occurring in the first 60 days of the mission. The trend was observed both in the calibration data and in the sky data. Therefore, an exponential drift of the ICAL thermometer was included in the fit to remove this trend, which was seen in all of the channels. The time constants found were 60–100 days depending on the channel. The source of this trend was not determined, although it was clearly in the instrument since it was observed both with the XCAL in and out of the horn. The sky data were used to remove it since those data have a larger signal-to-noise ratio than the smaller calibration data set.

The complete calibration model consisted of 2397 free parameters for the low-frequency channels and 4293 for the high-frequency channels (e.g. Ref. 25). Using only the calibration data which were taken near the null condition, the χ^2 per degree of freedom is close to 1.0. The internal consistency of the gain in the high-frequency channels is 3%, which is also consistent with the DIRBE photometry at 240 μm.

From the internal consistency of the calibration, we find that the calibration data are good to less than 0.1% of the peak of the CMBR for relative measurements. The dominant systematic error in the high frequency channels is a gain uncertainty which is less than about 3%. The limits on most individual pixels are set by the noise of the detector. The absolute temperature accuracy is limited by the error on the XCAL thermometry.

6.7 FREQUENCY SCALE CALIBRATION

The frequency calibration of the FIRAS was determined from observations of the interstellar [C II] and [N II] emission lines. Recent analysis of these lines in the high-frequency spectra has shown that the frequency calibration is self-consistent to 0.084% at 60 cm^{-1} (1799 GHz). Analysis of the CO lines in the low frequency spectra has also shown that the frequency calibration is self-consistent to 0.152% at 15 cm^{-1} (450 GHz). The difference in the frequency scale errors at 15 and 60 cm^{-1} may arise from beam divergence within the instrument. The frequency calibration is also consistent with measurements of the positions of all the other detected interstellar lines, which will be discussed in Section 7.2 .

6.8 TEMPERATURE SCALE CALIBRATION

The temperature scale of the FIRAS was calibrated in two ways. In the first method, temperature readings from two of the three GRTs attached to the base ring on the XCAL were averaged to produce a measured temperature of the calibrator. In the second method, the calibration model included an additive correction to the measured temperature of the XCAL external calibrator. This parameter allowed us to determine the color temperature scale of the instrument, which ties the calibrated frequency scale to temperatures through the form of the Planck function. This topic will be discussed in more detail in Section 7.3.

6.9 DESTRIPING

The FIRAS destriper is a secondary calibration program that corrects for residual instrumental effects in the calibrated spectra. It assumes that the calibration model does not sufficiently account for all the 'sources' in the FIRAS instrument; and that these effects can be removed, to some extent, by offset (subtractive) correction spectra. The most prominent of these are a smoothly decreasing temporal feature particularly noticeable during the beginning of the mission and a relatively sharp jump observed in spectra taken during the Earth eclipse season when the sky and reference horn temperatures were raised to 4 and 6 K, referred to as the 'hot horns' periods. Unlike the primary calibration program, the destriper uses both sky and calibration data, minimizing the discrepancy between multiple observations within the same sky pixel for the former and deviations from a Planck (blackbody) spectrum for the latter.

The destriper minimizes the function

$$\chi^2 = \sum_i \left[S_i^{(\mathrm{sky})} - A_{p(i)} - \sum_k J_k f_{ki} \right]^2 n_i + \sum_i \left[S_i^{(\mathrm{cal})} - P_i - \sum_k J_k f_{ki} \right]^2 n_i \qquad (8)$$

with respect to A_p, the destriped spectrum for the pth sky pixel and J_k, the offset spectra for the kth correction. These quantities are computed independently for each frequency so we have suppressed the frequency index. $S_i^{(\mathrm{sky})}$ is the ith calibrated sky spectrum, $S_i^{(\mathrm{cal})}$ is the ith calibrated calibration spectrum, P_i is a Planck spectrum at the XCAL temperature of the ith calibration observation, f_{ki} is the value of the kth model function for the ith sky or calibration observation and n_i is the corresponding effective weight for this observation. It is essentially the number of interferograms used to form a spectrum, adjusted for the effects of glitches. This correction 'de-weights' co-added spectra with higher glitch rates relative to those with lower rates.

The smooth temporal feature was modeled with two exponentials of the form $\exp(-t_i/\tau)$ with $\tau = 61$ and 153 days, where t_i is the time of the ith observation in days from aperture cover ejection. The values for τ are optimized for the 'left–low : short–slow' channel but applied to all other channels and scan modes. The hot horns are described by two 'top hat' functions that are unity if an observation was made during a hot horn period and zero otherwise. The fifth and final model function is an entire mission offset that is unity for all observations.

A linear fit was made to the sky and calibration co-adds from all four low-frequency channels at each frequency. The fit includes 14 time-dependent templates, four temperature templates and a parameter for the sky emission at each pixel and frequency. Thus 18 stripe parameters and 6068 fluxes are used to fit ~100 000 co-adds separately at each frequency. By simultaneously fitting all of the data sets with a single 'best' sky and a photometric model, the maximum amount of cross-checking on each data set was obtained. By simultaneously fitting the calibration data the data are effectively recalibrated, the best sky signal is obtained and a covariance matrix can be derived to measure the correlations introduced by this process.

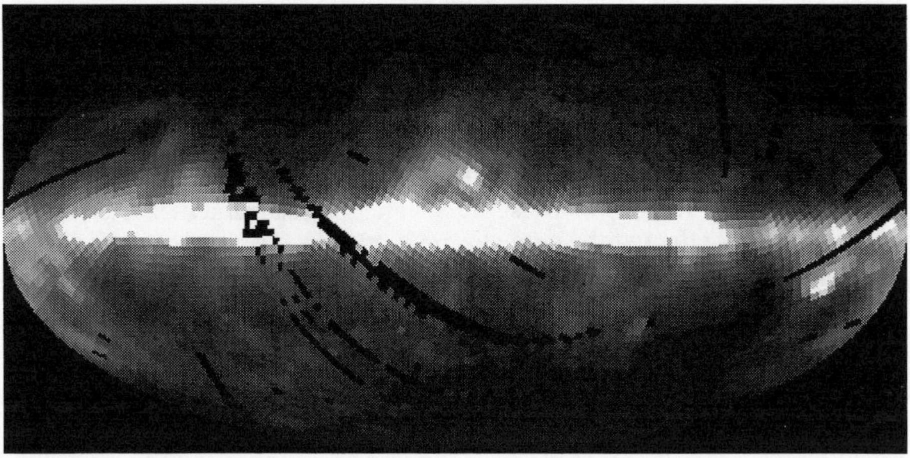

Figure 6 FIRAS far-infrared sky map showing the Galactic plane and emission at high Galactic latitudes

7 SCIENTIFIC RESULTS

At the wavelengths to which FIRAS is most sensitive, the sky presents itself as a complex patina of far-infrared and microwave sources which is revealed by the far-infrared observations presented in Figure 6. The most significant of these is the Milky Way, which is seen via the thermal emission of dust grains found in interstellar clouds. Most of this emission occurs within the plane of the Milky Way, but to some degree the rest of the sky is also cluttered by the so-called infrared 'cirrus' clouds, which are small, filamentary nearby clouds.

7.1 MODELING GALACTIC DUST

The detailed analysis of the FIRAS long-wavelength spectra required fitting models. For each pixel and spectrum, there were at least four physical contributions in the sky: a monopole term, $I_0(\nu)$, a dipole contribution, $D(l, b)$, due to the Doppler effect, where l and b are the Galactic longitude and latitude, and two contributions by the Milky Way, $G_1(l, b)$ and $G_2(l, b)$. The calibrated destriped sky spectra were then fitted to these four spatial and spectral templates:

$$S(\nu; l, b) = I_0(\nu) + D(l, b)d(\nu) + G_1(l, b)g_1(\nu) + G_2(l, b)g_2(\nu) \qquad (9)$$

The fit was made independently at each frequency. Only the spatial variation was assumed. To make this separation, the functions $D(l, b)$ for the dipole and $G_k(l, b)$ for the Galactic emission must be specified. The dipole is

$D(l, b) = \cos\theta$, where θ is the angle between the observation and the maximum of the dipole, $(l, b) = (264.26°, +48.22°)$ [26].

For $G_k(l, b)$, several combinations of predictors were tried including a plane-parallel, $\csc|b|$ distribution, the spatial distribution of the total power received in the high-frequency FIRAS channel above 25 cm^{-1}, and various forms of the DIRBE data. For the Galaxy templates, we use the normalization $\langle G(l, b)\rangle_{|b|>60°} = 1.074$ which is the natural normalization of the $\csc|b|$ model. The only effect of the normalization, of course, is to rescale the Galaxy spectrum. With this normalization, the resulting Galaxy spectrum is the sky brightness near the Galactic poles.

The Galactic spectrum can be approximately fitted by functions of the form $\nu^n B_\nu(T_{\text{dust}})$. Wright *et al.* [7] examined the FIRAS all-sky average Galaxy spectrum and found that it was fitted by $n = 1.65$ and $T_{\text{dust}} = 23.3$ K. The Galactic spectrum derived from high-latitude Galactic data produces a much lower temperature, $T_{\text{dust}} = 13 \pm 1$ K, with n fixed at 2 and an emissivity of $\sim 10^{-4}$ at 60 cm^{-1}. With only the low-frequency data used, fits for both n and the temperature were not possible. About half of the temperature difference is due to the change in index from 1.65 to 2 and half from the change to the higher latitudes.

The Galactic fit to $\csc|b|$ has more radiation at the Galactic poles than the fit to the other Galactic models, indicating that the Sun is in a local 'hole' in the Galaxy. This local 'hole' is approximately two-thirds as dense as a typical Galactic location.

The dust can also be modeled with other forms. Most of the information on the Galactic regions is at higher frequencies. A full discussion of the Galaxy requires the high-frequency data [27, 28]; here the Galactic radiation is treated as a contaminant of the CMBR.

7.2 GALACTIC LINE EMISSION

In addition to continuum emission from dust, most of the emission lines listed in Table 3 were also detected by FIRAS from specific gaseous ingredients such as carbon monoxide and atomic forms of carbon, nitrogen and oxygen which are also prominent as a consequence of their role in the cooling of interstellar clouds (e.g. Ref. 27). Among the lines detected were CO, [C I], [C II] and [N II], of which the [C II] and [N II] lines at 158 and 205.3 μm were strong enough to be mapped outside the Galactic plane. The [C II] line arises predominantly from photodissociation regions and the [N II] line from the diffuse warm interstellar medium.

Note that the CO $J = 1-0$, CO $J = 6-5$, O_2, H_2O, [O I] and [Si I] lines have not been definitively detected by FIRAS, though the detection of the 37.136 cm^{-1} H_2O line in absorption at the Galactic center is suggested by the data.

Table 3 Spectral lines in FIRAS data

Species	ν (cm^{-1})	ν (GHz)	λ (μm)
CO	3.845	115.27	2600.7
CO	7.690	230.54	1300.4
CO	11.535	345.80	867.0
O_2	14.168	424.75	705.8
CO	15.378	461.04	650.3
C I	16.419	492.23	609.1
H_2O	18.576	556.89	538.3
CO	19.222	576.27	520.2
CO	23.065	691.47	433.6
C I	27.00	809.44	370.4
H_2O	37.136	1113.3	269.3
N II	48.738	1461.1	205.2
C II	63.395	1900.5	157.7
O I	68.716	2060.1	145.5
Si I	77.11	2311.7	129.7
N II	82.036	2459.4	121.9

To determine and map the line fluxes, a separate line profile model was developed. The line profile model assumes that the interferogram of each line is a simple cosine subject to self-apodization arising from beam divergence effects. The measured self-apodization of the instrument at the 157.7 μm [C II] line is best described by a Gaussian with a width of 156 sample points. The model line interferogram is:

$$I(x) = \cos(2\pi\nu x)\exp\left[-\frac{1}{2}\left(\frac{\nu}{\nu_0}\frac{x}{\sigma}\right)^2\right] \tag{10}$$

where x is the optical path difference, ν is the line frequency, ν_0 is the Nyquist frequency of the instrument (144.981 cm^{-1} or 4346.4 GHz for low-resolution data, 36.245 cm^{-1} or 1086.6 GHz for high-resolution data) and σ is the width of the Gaussian self-apodization function at the Nyquist frequency.

The synthetic interferogram is processed in a manner similar to that used to generate sky spectra from co-added IFGs. The interferogram is further apodized with the apodization function described in Section 7.2, then Fourier transformed to produce a complex-valued, uncalibrated synthetic single-line spectrum. The normalization of the transform gives the spectrum the units of GHz^{-1}. At this point, the processing of the synthetic spectrum diverges from that of the sky spectra. The synthetic spectrum is 'calibrated' by multiplying the spectrum on a pointwise basis by the gain function for the appropriate scan mode. The spectrum is then divided at each data point by the value of the gain function at the frequency of the line. The resulting calibrated synthetic line spectrum or line profile has the units of GHz $^{-1}$:

$$S(\nu') = \Omega^{-1} \frac{G(\nu')}{G(\nu)} \int_{-\infty}^{+\infty} \mathrm{d}x \exp(2\pi i\nu' x) A(x) \cos(2\pi\nu x) \exp\left[-\frac{1}{2}\left(\frac{\nu}{\nu_0}\frac{x}{\sigma}\right)^2\right] \quad (11)$$

where ν' is the frequency along the line profile, ν is the line frequency, Ω is the normalization of the transform, G is the gain function and $A(x)$ is the apodization function.

7.3 COSMIC BACKGROUND SPECTRUM

Since the 1930s, the premier paradigm for cosmic evolution has been Big Bang cosmology, more technically Friedman–Robertson–Walker cosmology, which uses Albert Einstein's theory of general relativity to describe the observed large-scale motion of galaxies. Gamow [29] and Alpher and Herman [30] predicted that such a universe would have a very hot initial phase in which fusion among protons and neutrons would create hydrogen, deuterium and helium. Alpher and Herman also deduced that this early phase would be dominated by a background of electromagnetic energy at temperatures of many billions of degrees in thermal equilibrium with matter. As the universe expanded, this 'fireball' radiation would be red shifted to lower frequencies through the optical and on into the sub-millimeter and radio range. This prediction went largely unappreciated until 1965, when Arno Penzias and Robert Wilson at Bell Laboratories detected the 'hiss' from this cosmic microwave background radiation (CMBR) at 7.35 cm while investigating the nature of the noise in their telecommunication electronics [1]. The temperature was found to be near 3 K and surprisingly isotropic (uniform) in the sky, consistent with the earlier predictions by Alpher, Herman and Gamow using the Big Bang theory. Within a few weeks of operation, the FIRAS had confirmed that the CMBR was virtually indistinguishable from a perfect Planck distribution even with less than 9 min of accumulated interferograms. These results, shown in Figure 1, were first announced at an American Astronomical Society meeting in Washington, DC, on 13 January 1990.

There are three ways of determining the CMBR temperature from the FIRAS data set: (1) the preflight calibration of the external calibrator thermometers could be used, which should be good to the nominal 1 mK accuracy of the calibration specification; (2) the calibration data could be used to determine the temperature scale; or (3) the CMBR itself could be used if it is assumed that the dipole is a result of a Doppler shift. The details below are as given by Fixsen *et al.* [6, 8].

Method 1 is simple since there exist substantial calibration data with the external calibrator at 2.73 K and other temperatures near the sky conditions. This gives a CMBR temperature of 2.730 ± 0.001 K, with the error entirely dominated by the absolute thermometry calibration error on the external calibrator. The main difficulty with this is that there is evidence that the internal calibrator thermometry has errors of ~3 mK and that the thermometer at the tip of the external calibrator differs from the other two by 3 mK (at 2.73 K). There is some evidence from testing the alternate calibrators on the ground that both of these problems may be due to insufficient thermal contact for the thermometers embedded in epoxy at the tip, and that the two thermometers on the external calibrator fixed to the copper base ring should not have this problem. These were used in estimating the external calibrator temperature and the tip temperature is not used.

Method 2 effectively uses the 'color' to determine the temperature and so depends on the frequency determination. While Mather *et al.* [4] and Fixsen *et al.* [24] use the [C II] (63 cm^{-1}) and [N II] (49 cm^{-1}) lines, the possibility exists that the high frequency calibration and the low-frequency calibration need not agree. A discrepancy can arise because the interferometer has 14° beam divergence which, because of the cosine effect, implies a frequency shift. However, the beam profile of the high-frequency light may be more strongly attenuated than the beam profile of the low-frequency data. To improve the frequency scale we used a combination of CO and [C I] lines (7.69, 11.53, 15.38 and 16.42 cm^{-1}) to make the low-frequency determination to ±0.03%. The difference between the high- and low-frequency scales implies a difference of 1° in average beam divergence. The 0.03% frequency uncertainty implies a temperature uncertainty of 0.82 mK. There is an additional 0.2 mK error in determining the color temperature once the frequency scale is set, but as this adds in quadrature it is negligible. The result of this analysis is that the CMBR temperature is 2.7255 ± 0.0009 K.

Finally, the third method is based on the fact that the shape of the differential spectrum should be dB_ν/dT, where $B_\nu(T)$ is the Planck function. The best fit temperature to the dipole spectrum (see below) is a CMBR temperature of 2.717 ± 0.007 K. This result depends on a differential spectrum so offsets of the instrument calibration do not matter. Since the measurements are taken 50 min apart (on opposite sides of the orbit), long-term drifts (of the sort we see and correct for) also do not matter. This measurement is not completely independent because it still depends on the frequency calibration. The uncertainty is dominated by the uncertainty in fitting the Galaxy radiation which modulates the dipole signal, which is itself only 0.1% of the CMBR signal in the Rayleigh–Jeans region.

These methods give answers that are within 3σ of one another. Apparently there are additional systematic errors in one or more of these methods. Since the color temperature and the thermometer calibration are independent, the weighted sum and the weighted uncertainty may be computed for which the nominal uncertainty is 0.65 mK. However, in view of the history of error estimation and the unmodeled errors that might be present, it was concluded that the absolute temperature of the CMBR is 2.728 ± 0.004 K (95% confidence level), entirely dominated by the systematic errors.

A detailed fitting of the FIRAS spectra to the multicomponent model given by equation 9 indicated that the CMBR monopole spectrum is well fitted by a Planck blackbody spectrum. The deviations are small, consistent with the earlier FIRAS results within their larger uncertainties [4, 5, 25, 31]. To determine or constrain any deviations from a blackbody, consider a generic cosmological model $S_c(\nu; p)$, where p is some cosmic parameter quantifying the deviation from a blackbody, such as the Kompaneets [32] y parameter for Comptonized spectra or the dimensionless chemical potential μ for a Bose–Einstein photon distribution. Because the deviation is small, a linear fit

$$I_0(\nu) = B_\nu(T_0) + \Delta T \frac{\partial B_\nu}{\partial T} + G_0 g(\nu) + p \frac{\partial S_c}{\partial p} \qquad (12)$$

can be performed to the unknown parameters p, G_0 and ΔT. The first two terms are the Planck blackbody spectrum, with the temperature $T_0 + \Delta T$. The third term allows for Galactic contamination remaining in the monopole spectrum. The final term is the modeled deviation. The Kompaneets parameter or the chemical potential can be fitted individually, but the two are too similar to fit them simultaneously. The uncertainties are propagated from the template fits, and the correlation between $g(\nu)$ and $\partial S_c/\partial p$ increases the uncertainty of G_0 and p.

The determination of $g(\nu)$ is dominated by low-latitude Galactic emission, and there is some variation from this form at higher latitudes. The Galactic latitude cut-off used in deriving $I_0(\nu)$ can be varied to test the effect of variations in the Galactic spectrum from $g(\nu)$. Variations greater than the statistical uncertainty in any derived parameters, such as the cosmological term p, would most likely be due to an inadequacy in our Galactic model. Indeed, if the χ^2 cut-off is raised and the csc $|b|$ model is used, such effects can be seen.

When p is fixed at 0, i.e. there is only a Galactic deviation fit, the weighted rms deviation in the frequency range 2–21 cm^{-1} is 50 ppm of the peak brightness. The weighted rms is defined as

$$\text{rms}_w = \sqrt{\sum_\nu \left(\frac{\Delta_\nu}{\sigma_\nu}\right)^2 \Big/ \sum_\nu \frac{1}{\sigma_\nu^2}} \qquad (13)$$

where Δ_ν is the difference of the model and the data at spectrum point ν, and σ_ν is the uncertainty of the spectrum point ν. The formal χ^2 per degree of freedom $= 46/40 = 1.15$. The best fit temperature is 2.728 K, which is determined by the choice of the frequency and thermometry scales. When comparing this with values from other experiments, the systematic thermometry uncertainty of 2 mK (1σ) should be included. The detector noise part of the temperature uncertainty is only 10 μK, and is entirely negligible.

There are two likely forms for distortions of the primeval spectrum produced during the plasma epoch before recombination. Energy release or conversion in the red-shift range $10^5 < z < 3 \times 10^6$ produces a Bose–Einstein distribution, where the Planck law is modified by a dimensionless chemical potential μ [32]

$$S_\mu(\nu; T, \mu) = \frac{2hc^2\nu^3}{e^{x+\mu} - 1} \tag{14}$$

where $x = hc\nu/kT$ and ν is measured in cm^{-1}. The current FIRAS result is $\mu = -1 \pm 4 \times 10^{-5}$, or a 95% confidence level upper limit of $|\mu| < 9 \times 10^{-5}$.

Energy release at later times, $z < 10^5$, produces a Comptonized spectrum, a mixture of blackbodies at a range of temperatures. In the case of non-relativistic electron temperatures this spectrum is described by the Kompaneets equation [32], parameterized by the value of y [34]:

$$y = \int \frac{k(T_e - T_\gamma)}{m_e c^2} \, d\tau_e \tag{15}$$

where T_e, T_γ and τ_e are the electron temperature, the CMBR photon temperature and the optical depth to electron Compton scattering, respectively. The distortion will be of the form [34]

$$\frac{\partial S_y}{\partial y} = T_0[x \coth(x/2) - 4] \frac{\partial B_\nu}{\partial T} \tag{16}$$

The result is $y = -1 \pm 6 \times 10^{-6}$. There is some dependence on Galactic cut and on the Galactic model. This indicates either residual Galactic contamination or a nearly uniform source with a spectrum similar to the Galactic spectrum. A systematic uncertainty of 4×10^{-6} is included to account for these effects for a final 95% confidence level upper limit of 15×10^{-6}.

The limit for $|\mu|$ is about one-third as large as that given in Mather *et al.* [4] and the limit on $|y|$ is about half as large. The $|y|$ is more strongly coupled to the residual Galactic contamination. These new values place stringent constraints on theories of the early universe and the development of cosmic structure.

7.4 *DIPOLE SPECTRUM*

The dipole spectrum was fit to the derivative of a Planck function $T_{amp}\, dB_\nu(T_{cmbr})/dT$, where $T_{cmbr} = 2.728$ K was fixed and T_{amp} allowed to vary. The best-fit value of the dipole amplitude shown in Figure 7, is $T_{amp} = 3.369 \pm 0.004$ mK. However, the value is dependent on the Galactic cut-off and the Galaxy model.

In the 'cloud' of solutions made with different assumptions about the

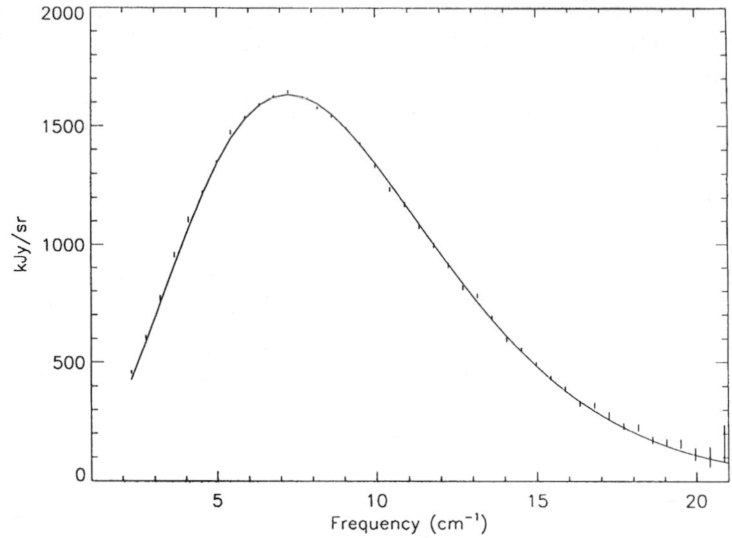

Figure 7 FIRAS dipole spectrum with $\pm 1\sigma$ error bars

Galaxy, an rms variation of 0.006 mK was found. Adding this in quadrature gives an uncertainty of 0.007 mK. A 0.003 mK correction was also made for the finite beam size, which is approximately 7° FWHM. This FIRAS-determined dipole amplitude agrees with 3.353 ± 0.024 mK measured by the *COBE* DMR [26], derived from observations at 1.05, 1.77 and 3.00 cm^{-1}. The DMR beam is also approximately 7° but the details of the beam patterns are different. Possible effects of the octupole and higher harmonics are not included, although because of the non-uniform sky coverage by FIRAS, these might perturb the final result.

By choosing the monopole temperature as the point to evaluate $\mathrm{d}B_\nu/\mathrm{d}T$, the dipole temperature was forced to be that of the monopole. If both T_{amp} and T_{cmbr} are fitted, FIRAS obtains $T_{\mathrm{cmbr}} = 2.717 \pm 0.003$ K (1σ statistical uncertainty). However, the result is sensitive to the Galactic cut, which contributes an uncertainty of 0.006 K. This results in a final estimation of $T_{\mathrm{cmbr}} = 2.717 \pm 0.007$ K (1σ combined uncertainty). The agreement of the dipole temperature with the monopole temperature is consistent with the conventional Doppler interpretation of the dipole. It is also confirmation that the overall temperature scale and calibration for the FIRAS are consistent.

The dipole direction from the FIRAS data can also be determined from a different fitting procedure. The vector sum of the dipole coefficients points in the direction $(l, b) = (264.14° \pm 0.15,\ 48.26° \pm 0.15)$, consistent with the direction from the DMR results.

8 INTERPRETATION

The FIRAS spectrum of the cosmic microwave background radiation agrees with a blackbody spectrum to high accuracy. The CMBR dipole has a spectrum consistent with its thermal origin and a Doppler shift. The dipole itself has a differential thermal spectrum the temperature of which, 2.717 ± 0.007 K, agrees with the monopole temperature. The Doppler shift implies that the Sun's peculiar velocity relative to the comoving frame is 371 ± 1 km s^{-1} (95% confidence limit) toward $(l, b) = (264.14° \pm 0.15, 48.26° \pm 0.15)$, in agreement with the microwave results from the DMR.

The CMBR temperature is 2.728 ± 0.004 K (95% confidence limit), where the error is dominated by the estimate of the thermometry errors. The weighted rms deviation from the fit is 50 ppm of the peak brightness. The limit on $|y|$ is 15×10^{-6} and the limit on $|\mu|$ is 9×10^{-5} (95% confidence limit).

The cosmological interpretation of the limits on y and μ was developed by Wright *et al.* [7]. Since the FIRAS spectrum with the Galactic dust emission removed is close to a blackbody, strict limits can be set on energy release after the Big Bang. For energy release during the Bose–Einstein period, the limit is $\Delta U/U = 0.71\mu$, where U is the energy of the cosmic background and ΔU is the energy converted from other forms. For energy release after that but before decoupling, the limit is $\Delta U/U = 4y$. The result can be summarized as a limit on $\Delta U/U$ as a function of red shift. The implications of these upper limits are that <0.006% of the energy in the cosmic microwave background was added to it after the first year of the expansion of the universe, and that <0.6% of the primordial hydrogen could have been burned by any first generation Population III stars after a red shift of 80. The FIRAS results, when taken together with the other COBE findings, provide overwhelming support for the standard Big Bang scenario. They agree with, but do not prove, a slight modification to the physics of the expansion during an 'Inflationary' phase that occurred about 10^{-34} s after the Big Bang. This proposed modification allows structure to exist in the universe which was much larger than the horizon at any epoch, and also explains the large linear scale of the inhomogeneities seen by the COBE diffuse microwave radiometer [19, 26]. There was no evidence, however, that at any time after the end of the Nucleosynthesis Era, some 10 min after the Big Bang, significant new sources of entropy (photons) were available. According to the FIRAS, the current ratio of photons to baryons in the universe of 10^8 has evidently not been altered since the end of the first year of cosmic expansion.

9 FUTURE OPPORTUNITIES

COBE/FIRAS demonstrated that even an instrument as complex as a cryogenic Michelson polarizing interferometer could be flown into space and remotely operated to return high-quality data. The FIRAS results are an order of magnitude more accurate than originally expected, based largely on software corrections for systematic errors. Given sufficient scientific interest, it would be possible to develop a considerably more accurate apparatus. New bolometer detectors operating at temperatures of the order of 50 mK are available that are two to three orders of magnitude more sensitive than those used in the FIRAS. Assuming that the cosmic ray glitches could all be recognized and removed, the sensitivity achieved over the course of 10 months of observation could now be obtained in less than one hour. With the use of a parabolic reflector instead of a concentrator, much finer angular resolution could be achieved. This would allow observations of the interstellar dust in great detail. The dust emission already limits the accuracy of the CMBR spectrum measurements, but some improvements would be expected, especially considering that the distribution of the infrared cirrus clouds is anisotropic and large areas of low cirrus emission exist. An improved FIRAS could also have redundant external calibrators, better immunity to thermal gradients and a more exact differential symmetry.

10 ACKNOWLEDGMENTS

The National Aeronautics and Space Administration/Goddard Space Flight Center (NASA/GSFC) is responsible for the design, development and operation of the Cosmic Background Explorer (COBE), under the scientific guidance of the COBE Science Working Group. GSFC is also responsible for the software development and the final processing of the mission data. The COBE mission is sponsored by the Astrophysics Division of NASA Headquarters. I especially thank the members of the FIRAS team who made this instrument such a success. Robert Maichle and Michael Roberto led the engineering team, and Richard Isaacman, Shirley Read, Eugene Eplee, Jr., Alice Trenholme and Stefan Brodd led the software teams. Richard Shafer serves as Deputy Principal Investigator for the FIRAS, and Dale Fixsen developed the FIRAS calibration software. Sten Odenwald served as the able and diligent editor of the manuscript.

11 REFERENCES

1. A.A. Penzias and R.W. Wilson, *Astrophys. J.*, **142**, 419 (1965).
2. J. Mather, D. Fixsen and R. Shafer, *Proc. SPIE*, **2019**, (1993).
3. COBE, in *FIRAS Explanatory Supplement*, Eds J.C. Mather *et al.*, National Space Science Data Center, Washington, DC (1995); available by anonymous FTP from nssdca.gsfc.nasa.gov.
4. J. Mather, *et al.*, *Astrophys. J.*, **420**, 440 (1994).

5. J. Mather, *et al.*, *Astrophys. J. (Lett.)*, **354**, L37 (1990).
6. D.J. Fixsen, *et al.*, *Astrophys. J.*, **420**, 445 (1994).
7. E.L. Wright, *et al.*, *Astrophys. J.*, **420**, 450 (1994).
8. D.J. Fixsen, *et al.*, *Astrophys J.*, in press.
9. P. Fellgett, *J. Phys. Radium*, **19**, 187, 237 (1958).
10. P. Jacquinot, *J. Phys. Radium*, **19**, 223 (1958).
11. H.A. Gebbie, *J. Phys. Radium*, **19**, 230 (1958).
12. E. R. Peck, *J. Opt. Soc. Am.*, *38*, 66 (1948).
13. H.H. Hopkins, *Opt. Acta*, **2**, 23 (1955).
14. D. Martin and E. Puplett, *Infrared Phys*. **10**, 105 (1969).
15. D.P. Woody, J.C. Mather, N.S. Nishioka and P.L. Richards, *Phys. Rev. Lett.*, **34**, 1036 (1975).
16. D.P. Woody and P.L. Richards, *Phys. Rev. Lett.*, **42**, 925 (1979).
17. D.P. Woody and P.L. Richards, *Astrophys*. J., **248**, 18 (1981).
18. N. Boggess, *et al.*, *Astrophys. J.*, **397**, 420 (1992).
19. G. Smoot, *et al.*, *Astrophys. J. (Lett.)*, **396**, L1 (1992).
20. M. Hauser, in *Unveiling the Cosmic Background Radiation*, ed. E. Dwek, AIP Conference Proceedings Vol. 348, pp. 11–24, AIP, New York (1996).
21. H. Hemmati, J. C. Mather and W. L. Eichhorn, *Appl. Opt.*, **24**, 4489 (1985).
22. A. Serlemitsos, *Proc. SPIE*, **973**, 314 (1988).
23. J. Högbom, *Astron. Astrophys. Suppl.*, **15**, 417 (1974).
24. D.J. Fixsen, *et al.*, *Astrophys. J.*, **420**, 457 (1994).
25. E. Cheng, in *International Symposium on Observational Cosmology*, Eds G. Chincarini *et al.*, ASP Conference Series, Vol. 51, p. 503, ASP, San Francisco (1992).
26. C.L. Bennett, *et al.*, *Astrophys. J. (Lett.)*, **464**, L1 (1996).
27. C.L. Bennett, *et al.*, *Astrophys. J.*, **436**, 423 (1994).
28. W. Reach, *et al.*, *Astrophys. J.*, **451**, 188 (1995).
29. G. Gamow, *Phys. Rev.*, **70**, 572 (1948).
30. R. Alpher and R. Herman, *Nature (London)*, **162**, 774 (1948).
31. R. Shafer, *et al.*, *Bull. Am. Astron. Soc.*, **23**, 1398 (1992).
32. Ya.B. Zeldovich and R.A. Sunyaev, *APSS*, **7**, 20 (1970).
32. A.S. Kompaneets, *JETP*, **4**, 730 (1057).
34. Ya.B. Zeldovich and R.A. Sunyaev, *APSS*, **4**, 301 (1969).

3

Far-Infrared Interferometry by Using Synchrotron Radiation

TAKAO NANBA
Department of Physics, Faculty of Science, Kobe University, Kobe 657, Japan

and

MIKIHIKO IKEZAWA
Research Institute for Scientific Measurements, Tohoku University, Aoba-ku, Sendai 980, Japan

CONTENTS

Principles and Applications of Polarization-Division Interferometry
Edited by P. L. Polavarapu

1 INTRODUCTION

The commonest light source for far-infrared spectroscopy is a high-pressure mercury arc lamp, which has been used since the beginning of this century [1]. The spectral intensity distribution of the blackbody (BB) radiation drops steeply in an exponential manner on the long-wavelength side. The weak tail part of the radiation is used in the far-infrared region. Fourier transform interferometry has been developed as a spectroscopic technique to achieve a good signal to noise ratio, and the quality of low temperature detectors has been much improved [1, 2]. Efforts have been made to improve only the spectrophotometric methods and little development of far-infrared light sources has been achieved. In the millimeter wave region, many oscillators have been devised as the spectroscopic light sources [3]. Some of them have very high output intensities, but none of them has a completely continuous output spectrum over the whole millimeter region.

Synchrotron radiation (SR), which is emitted dominantly from an accelerated electron or positron beam at the storage ring, is a new type of light source which covers in principle a very wide energy region from the microwave to the X-ray region. SR emitted from electron beam in a curved path in a bending magnet of the storage ring is shown schematically in Figure 1. The acceptance angle (θ) in the horizontal plane is determined by the optics of the spectroscopic instrument used.

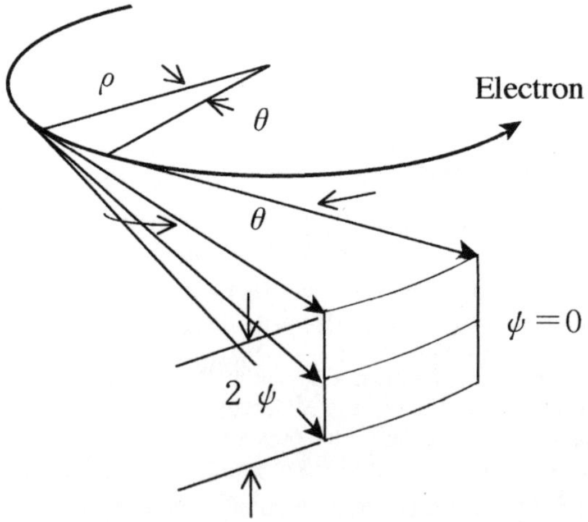

Figure 1 Schematic diagram of synchrotron radiation (SR) emitted from an accelerated electron beam in a dipole magnet with a bending radius ρ into a solid angle θ (horizontal direction) and ψ (vertical to the plane of the storage ring)

The emitted photon numbers $N_{k,\parallel}(\lambda, \psi)$ of SR for the parallel component $E_{\parallel}$ whose electric vector is parallel to the plane of the electron orbit and $N_{k,\perp}(\lambda, \psi)$ for the perpendicular component $E_{\perp}$ whose electric vector is perpendicular to the plane of the electron orbit per second, unit acceptance angle in the horizontal angle θ is expressed from the results of Schwinger [4] by

$$N_{k,\parallel}(\lambda, \psi) = 3.46 \times 10^6 k(\lambda_c/\lambda)^2 \gamma^2 (1 + X^2)^2 K_{2/3}(\xi) \quad (1)$$

and

$$N_{k,\perp}(\lambda, \psi) = 3.46 \times 10^6 k(\lambda_c/\lambda)^2 \gamma^2 (1 + X^2)^2 K_{1/3}(\xi) \quad (2)$$

where λ is the wavelength of the emitted photons (μm), k is a resolution factor which is equal to $\Delta\lambda/\lambda$, $X = \gamma\psi$, $\xi = (\lambda_c/2\lambda)(1 + \mathrm{X}^2)^{3/2}$, $\lambda = E/mc^2 = 1.957 \times 10^3 E$, E is an acceleration energy in GeV, ψ is the vertical angle between the direction of photon emission and the $\psi = 0$ plane of the ring, λ_c is the critical wavelength which is given by $\lambda_c = 4\pi\rho/3\gamma^3$, ρ is the radius of an electron orbit (m) in a circular motion and $K_{1/3}(\xi)$ and $K_{2/3}(\xi)$ are modified Bessel functions of the second kind. For the long-wavelength limit ($\xi \ll 1$), $K_{1/3}(\xi)$ becomes approximately proportional to $\xi^{-1/3}$ and $K_{2/3}(\xi)$ to $\xi^{-2/3}$.

In the early stages of the utilization of SR, it was used exclusively in the soft and hard X-ray regions [5, 6]. The fundamental properties of SR in the long-wavelength region have been experimentally examined be several groups [7–13]. As the light source for spectroscopy, SR has many excellent properties, such as high brilliance, a small divergence angle in the vertical direction to the plane of the ring, continuous spectral distribution in the intensity over the entire energy range and well defined pulsed properties. In particular, SR is a highly polarized light beam which has completely different characteristics to BB sources. For example, SR is a light beam whose electric vector contains mainly the parallel component ($E_{\parallel}$) and additionally a small amount of the perpendicular component (E_{λ}). According to the equation 2, the component $E_{\perp}$ of SR becomes zero in the $\psi = 0$ plane and then we obtain perfect linear polarization of $E_{\parallel}$. The parameters of the UVSOR ring are $E = 0.75$ GeV, $\rho = 2.2$ m and $\lambda_c = 29$ Å. Computed angular distributions of the SR from the UVSOR ring at different wavelengths in the far-infrared region are shown in Figure 2. The thin solid curves show the parallel component $E_{\parallel}$. The dashed curves show the component $E_{\perp}$ and the thick solid curves the summation. The beam diverges with wavelength in the vertical angle. For example, when we put a first collection mirror at a distance of 2000 mm from the curved part of the magnet, the extension of the light beam at the FMHW for the $E_{\parallel}$ component in the vertical direction amounts to 120 mm at a wavelength of 200 μm.

We can obtain almost linear polarization in the plane close to the $\psi = 0$ plane of the ring. Furthermore, we can obtain a light beam with circular polarization

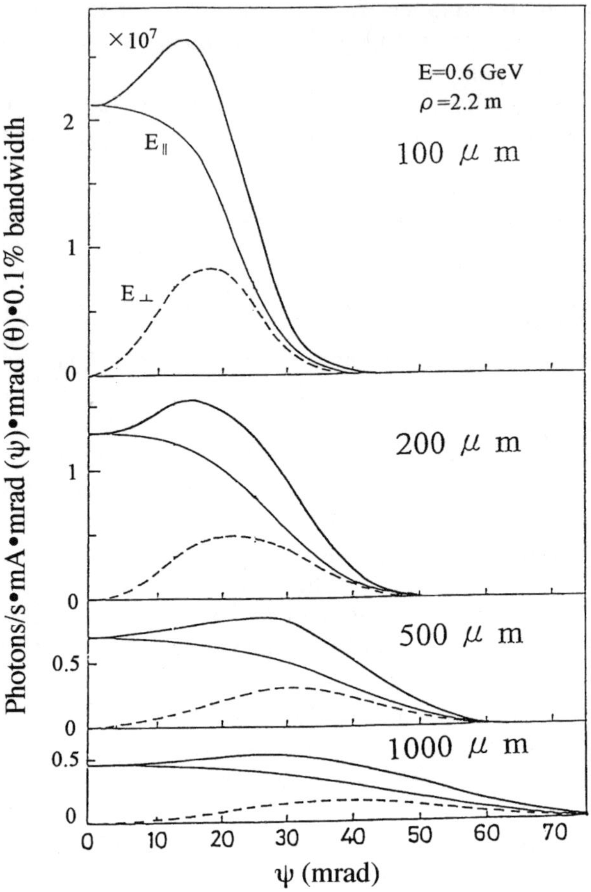

Figure 2 Computed light beam spread of UVSOR with an acceleration energy of 0.75 GeV, $\rho = 2.2$ m s^{-1}, horizontal angle (θ) as a function of vertical angle (ψ) at different wavelengths: (*a*) 100; (*b*) 200; (*c*) 500; and (*d*) 1000 μm. The thin solid curves correspond to the parallel components $E_{\parallel}$, the dashed curves to the perpendicular components $E_{\perp}$ and the thick solid curves to the sum

from SR without any polarizer at the off-axis plane of the ring. Such spectroscopy has been widely performed in the measurement of magnetic circular dichroism on magnetic materials using circular polarized SR in the X-ray to the vacuum ultraviolet (VUV) region. This is a prominent characteristic of SR, different from conventional light sources. We have the possibility of performing a circular dichroism experiment in the far-infrared region in the future.

Although the intensity of SR also drops gradually in the long-wavelength region, the decay in the intensity with wavelength is less steep than for BB radiation. Therefore, SR becomes superior in intensity to BB radiation for wavelengths longer than 100 μm in the far-infrared region. In the longer wavelength region compared with the X-ray and soft X-ray regions, SR diverges more and then one has to employ an optical system with a large acceptance angle to collect it efficiently [7–15]. In the far-infrared region, however, its brightness is still much higher than that of ordinary light sources such as a mercury lamp [9–12, 15]. Therefore, one can obtain a very narrow light beam for measurement using SR.

A spectroscopic system in the far-infrared region was completed at the beam line BL6A1 of the UVSOR, the electron storage ring facilities, of the Institute for Molecular Science at Okazaki in 1985 [12]. This chapter provides a detailed description of the present status of interferometric spectroscopy in the far-infrared region at UVSOR since its utilization began in 1985. SR has been confirmed to be extremely useful for spectroscopy in the long-wavelength regions, especially for solid state spectroscopy, owing to the excellent property of high brightness, among others. After long experience at UVSOR, SR was confirmed to be a convenient source for interferometric spectroscopy on small solid specimens and even for spectroscopy in the millimeter wave region. Measurements of the anisotropic reflectivity of small solid specimens of alkalimetal–silver and alkalimetal–cuprous halide crystals are described in the Section 4.1, and the reflection and transmission spectra of black phosphorus under pressure in Section 4.2. The millimeter wave absorption of germanium is described in the Section 4.3 as representing the extension of UVSOR to millimeter wave spectroscopy.

2 OPTICAL SYSTEM

2.1 OPTICAL CONFIGURATION

The optical configuration of the beamline BL6A1 is shown in Figure 3. The spectroscopic system consists mainly of a polarizing interferometer of Martin–Puplett type, which utilizes a highly polarized light beam (whose electric vector lies in the horizontal plane of the ring of the UVSOR ring) as the source, a mirror chamber to obtain a parallel light beam to the interferometer and a sample chamber.

The SR is led to the sample position by mirrors MP1, MS2 MP3, MP4, MT5, MP6, MT7 and MS8, where P, S and T represent a plane, spherical and toroidal mirror, respectively. All mirrors are made of glass coated with gold, except MP1, which is made of copper. The SR from the curved path in the dipole magnet is reflected by the plane mirror MP1 to the spherical mirror MS2. The plane mirror MP1 is not particularly cooled. MP1 and MS2 are at the same height as the light source. Mirror MP1 possesses a narrow slit (20 mm

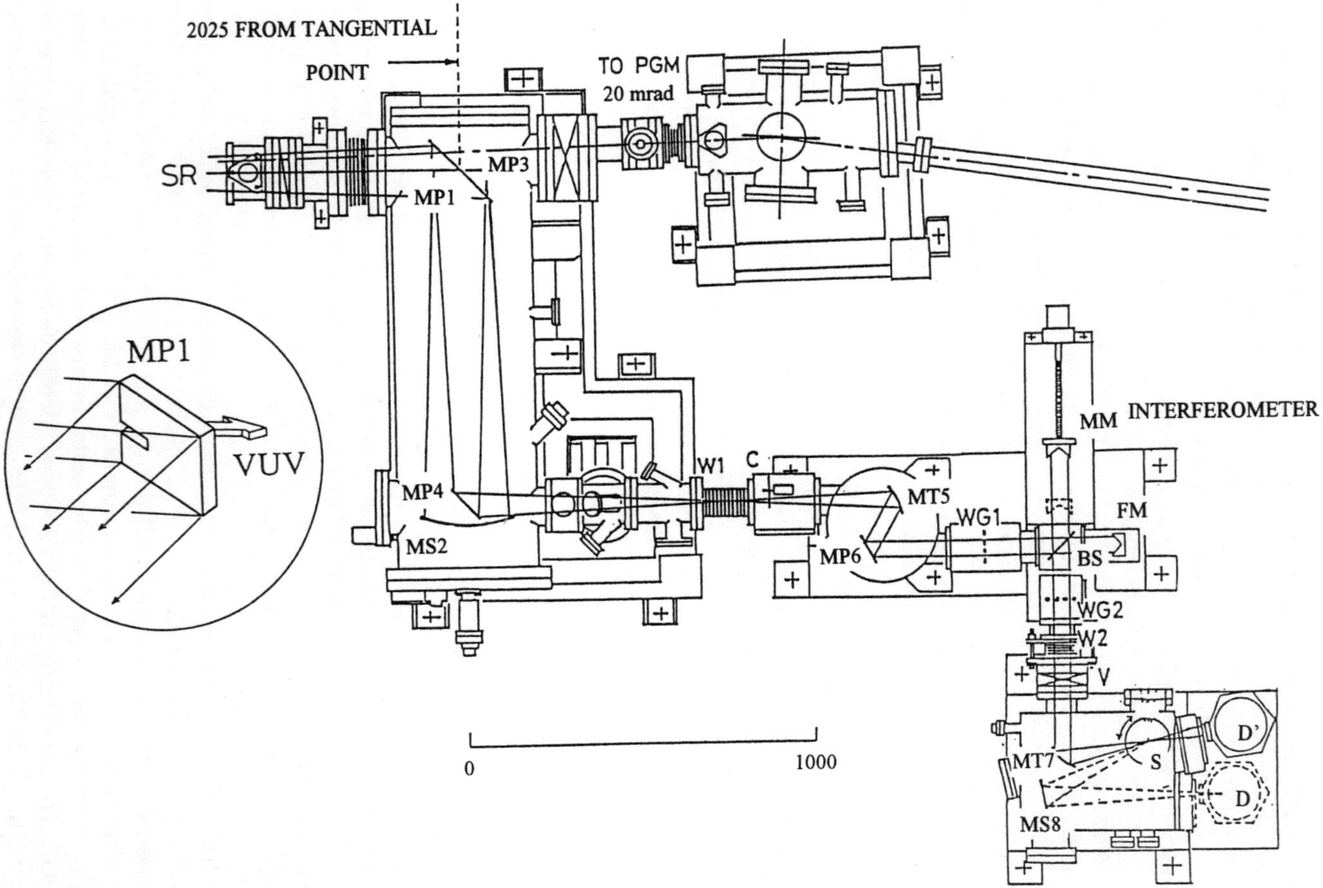
2025 FROM TANGENTIAL
POINT
TO PGM
20 mrad
SR
MP3
MP1
MP1
VUV
MP4
MS2
W1
C
MT5
MP6
WG1
MM
INTERFEROMETER
FM
BS
WG2
W2
V
D'
S
D
MT7
MS8
0
1000

width and 55 mm length), as shown in the inset in Figure 3. This is to supply the VUV component of the radiation to the beamline BL6A2 for VUV spectroscopy without a large decrease in the intensity of the far-infrared light. A small cut of the MP1 mirror supplies a light beam of 6 mrad for a vertical and 10 mrad for a horizontal acceptance angle to the BL6A2. Most of the infrared to far-infrared components are reflected by the MP1 mirror because of its large vertical divergence at UVSOR.

The light is reflected downward to a plane mirror MP3, which is at the level of the optical axis of the whole system downstream. After being reflected by another plane mirror MP3, the light beam is focused behind a wedge-shaped Si window W1. Silicon was selected as a optical window because of its flat transmissivity in the far-infrared region. The window separates the high-vacuum (10^{-7} Pa) part of the large mirror chamber from the low-vacuum (0.1 Pa) part of the interferometer. The effective diameter and the mean thickness of the window are 33 and 2 mm, respectively. The horizontal and vertical angles of acceptance of the collecting system by the MP1 are 80 and 60 mrad, respectively.

2.2 INTERFEROMETER

Since SR is a highly polarized beam, we adopted a polarizing interferometer (Specac, Orpington, UK) of the Martin–Puplett type (see Chapter 1). The light beam is collimated by the toroidal mirror MT5 and is reflected to the interferometer by a plane mirror MP6. The horizontal and vertical dimensions of the cross-section of the parallel beam are designed as 45×36 mm. The parallel or perpendicular component of the SR is selected by rotating the wire grid polarizer WG1 for the measurement of the degree of polarization of the SR beam. The spacing and the diameter of the wires of WG1 are 12.5 and 5 μm, respectively. This wire grid polarizer is usable over the whole spectral region. The wire grid WG1 is used also to select the SR or light from a mercury lamp by turning it by 90°. A 125 W mercury arc lamp is equipped with an interferometer as a reference source which is placed in the box above the incident

Figure 3 Plane view of the optical system for spectroscopy in the far-infrared and millimeter wave regions at the BL6A1 of the UVSOR ring. MP, MS and MT indicate plane, spherical and toroidal mirrors, respectively. MP1 and MS2 are at the same level as the orbital plane of the storage ring, which is higher than the level of the optical axis of the rest of the optical system. W1 and C are a Si window and a chopper, respectively. WG1 and WG2 are wire grid polarizers. FM and MM are fixed and moving mirrors, respectively. The sample is placed at S and the signal is detected by a detector at position D for reflectivity measurement or at D′ for transmission measurement. Inset: schematic diagram of mirror MP1 to show how to split the incident light beam partially to VUV spectroscopy. Scale in mm

polarizer box (not shown in Figure 3). The light from the lamp is collected with a parabolic mirror to form a parallel beam which travels to WG1. The solid angle of light collection is 0.075 sr.

The beamsplitter BS and the other polarizer WG2 in Figure 3 are wire grid polarizers with the same structure as WG1. The low-vacuum part of the interferometer is separated by a Mylar sheet of 16 μm thickness from the sample chamber at a high vacuum of 10^{-7} Pa.

The light is reflected and collected by the toroidal mirror MT7 at the sample position S. The vertical and horizontal angles of convergence of the light on to the specimen are 145 and 180 mrad, respectively. This convergence angle corresponds to an F-number of $F/6$, which is much larger than in a conventional spectroscopic system with a BB source. The horizontal or vertical component of the light is selected by the wire grid polarizer WG2 so that the light for measurement is linearly polarized. To measure the reflectivity, the reflected light from the specimen at S is collected by the spherical mirror MS8 in a detector at position D in Figure 3. In this case the angle of incidence is 7.5°. To measure the transmission of the specimen, the light from the specimen is collected by a copper horn at the position D′. A Ge bolometer (Infrared Laboratories, USA) was used as the detector. A double-sided interferogram is recorded, apodized with a quadratic cosine function and Fourier transformed to give a power spectrum using a microcomputer. The maximum scanning

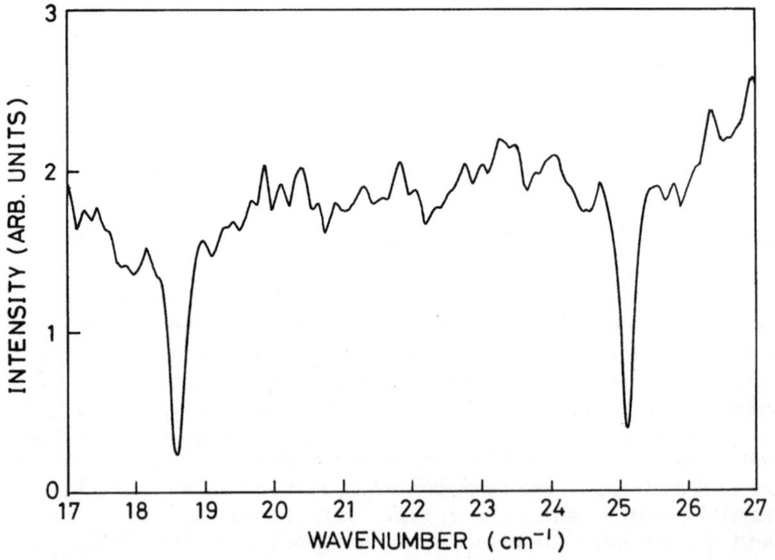

Figure 4 Absorption spectrum of H_2O vapor in the long-wavelength region. The nominal resolution was 0.1 cm^{-1}

distance is 38 cm and the minimum interval is 1.25 μm. Absorption measurements on water vapor were performed to examine the resolution $\Delta\omega$ of the spectroscopic system. The spectrum from 17 to 27 cm^{-1} is shown in Figure 4. A double-sided interferogram with a scanning distance of 10 cm^{-1}, which corresponds to $\Delta\omega = 0.1$ cm^{-1} after the apodization process, gave a spectrum with $\Delta\omega = 0.12$ cm^{-1}. This is sufficient resolution for solid-state spectroscopy at the BL6A1.

3 SPECTROSCOPIC CHARACTERISTICS OF OPTICAL SYSTEM

3.1 *BACKGROUND SPECTRUM AND BEAM SIZE AT SAMPLE POSITION*

For the purpose of measuring the cross-sectional size of the light beam, a diaphragm with variable diameter is placed at position S, which is the focal point of the incident light beam in the sample chamber. The diameter of the diaphragm is given in mm. The spectra were observed with a Golay cell at position D′ at different ring currents but their intensities were normalized to the spectrum of the current at 100 mA. The intensity of the light was confirmed to be proportional to the ring current at least up to 150 mA. The current of the stored electrons in the ring decreases with time, so the value of the current was recorded simultaneously with the interferogram to obtain a normalized signal intensity.

A series of observed spectra are shown in Figure 5. With a diameter of 10 mm the whole light beam passes through the diaphragm without any loss. The spectrum extends from 10 to 250 cm^{-1}. The high-energy side is lost, probably because of the low sensitivity of the Ge detector. The dips in the spectrum at about 76 and 131 cm^{-1} are due to absorption by polyethylene and quartz windows in the detector, respectively. The low-energy side of the spectrum is limited by the sensitivity of the detector and by the weak intensity distribution of the SR. An image of the light source is geometrically designed to focus at the sample position S in Figure 3. The size of the light source in the vertical direction is limited by the electron beam size, which is about 0.5 mm at most. From simple geometric considerations, the horizontal size is estimated as $\rho\theta^2/8$. Although the light intensity increases in proportion with the magnitude of ρ, a large ρ results in a large image at the sample position. In the present case, $\rho = 2.2$ m and $\theta = 80$ mrad, which give a horizontal size of 1.8 mm. Then the cross-sectional size of the light beam at the sample position is expected to be as small as 0.5×1.8 mm. When the diameter was made smaller, the intensity began to decrease at a diameter of about 3 mm on the high-energy side above 150 cm^{-1}. This result shows that the cross-section of the light beam

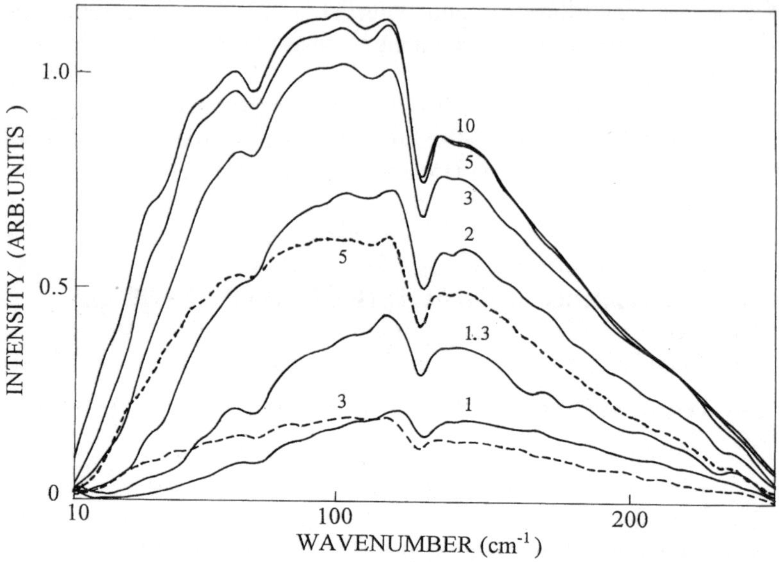

Figure 5 Spectral intensities in the far-infrared region detected by a Golay cell at D′ in Figure 3 and normalized to the ring current of 100 mA. The numbers on the spectra represent the diameter of the aperture at the sample position S through which the light beam passed. The dashed curves are the intensities of the mercury arc lamp through the apertures of 3 and 5 mm. The dip structures at 131 and 76 cm^{-1} correspond to absorption by crystalline quartz and a polyethylene filter, respectively

is as small as the diaphragm of 3 mm above 150 cm^{-1}. In the lower energy region, the intensity begins to decrease at a diameter of about 5 mm, which is larger than the geometrical]y calculated value. The extension of the cross-section of the beam at the sample position is probably due partly to the aberration of the optical system and partly to the diffraction effect.

The diffraction effect is also important in the measurement of small samples. To demonstrate this, the intensities through the diaphragm at wavenumbers of 150, 100 and 50 cm^{-1} are re-plotted versus the area of the diaphragm in Figure 6. The straight line is proportional to the area of the diaphragm. At 150 cm^{-1} the decrease in the intensity is proportional to the area of an aperture of 1 mm. At 100 cm^{-1} the decrease is steeper than the decrease in the area when the diaphragm is smaller than a few mm, and at 50 cm^{-1} the decrease is much steeper. This result indicates that the diffraction effect due to the diaphragm makes the light beam more divergent and decreases the detectable intensity when the wavelength is comparable to the size of the diaphragm.

In Figure 5, the intensities from the mercury arc lamp are shown by the dashed curves. The beam was received through diaphragms of 3 and 5 mm.

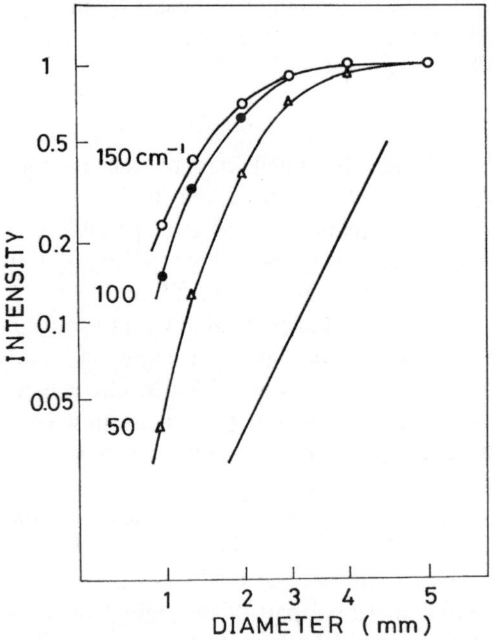

Figure 6 Intensities of light of 50, 100 and 150 cm^{-1} as a function of the diameter of the aperture at the sample position through which the light beam passed. The intensities are normalized to that obtained at the aperture of 10 mm. The straight line is proportional to the area of the aperture. The steep inclination in the intensity apart from the straight line means diffraction loss

The intensity of the SR at a wavelength of 100 μm was about four times that of a mercury lamp at a ring current of 100 mA when the diameter of the aperture was 5 mm. However, it can be seen that the intensity of the SR through the same aperture of 3 mm with a mercury lamp is about ten times more intense even in operation at a low current of 100 mA. This result shows that the main advantage of SR over the mercury lamp lies in the size of the measuring beam smaller than 8 mm in operation at a low current. At present, the operation of the UVSOR ring starts at a ring current of about 200 mA and consequently the real intensity is about 20 times more intense. The maximum current of the UVSOR ring is designed for 500 mA, at which the intensity of SR will be far greater than that of the mercury arc lamp at any size of the measuring beam.

3.2 DEGREE OF POLARIZATION AND DIFFRACTION EFFECT

The above observations were made with the horizontal component of the synchrotron radiation. By turning the wire grid polarizer WG1 by 90°, the spectrum of the vertical component was observed in the far-infrared region. Wire grid polarizers of fine spacing were used in this measurement. We define the degree of polarization as follows:

$$P = (I_{\parallel} - I_{\perp})/(I_{\parallel} + I_{\perp}) \quad (3)$$

where $I_{\parallel}$ and $I_{\perp}$ indicate the intensities of the parallel and perpendicular components, respectively, which are given by $I_{\parallel} = N_{k,\parallel}(\lambda, \psi)h\nu$ and $I_{\perp} = N_{k,\perp}(\lambda, \psi)h\nu$ by using the equations 1 and 2, respectively. Using equation 3, the degree of polarization of the source was calculated and is shown as the solid curve in Figure 7. In the high-energy region above about 200 cm^{-1}, the observed value is in good agreement with the calculated value with the designed acceptance angle of 60 mrad, i.e. $\psi = \pm 30$ mrad. On the lower energy side, however, the observed value is larger than the calculated value. This result may be interpreted by the effective acceptance angle decreasing with increase in wavelength because of the increase in the diffraction effect in the longer wavelength region as follows. As the area of the first plane mirror for the light collection is limited, it will cut the marginal part of the radiation from the source. Then the cross-section of the light beam diffracted by the first mirror will become larger than the surface of the next collection mirror, which was geometrically designed without considering the diffraction effect, and a marginal part of the light which contains more vertical component of the radiation will be lost. After repeated diffraction at subsequent mirrors, the effective acceptance angle of light collection will become smaller than 60 mrad in the long-wavelength region, as has been observed.

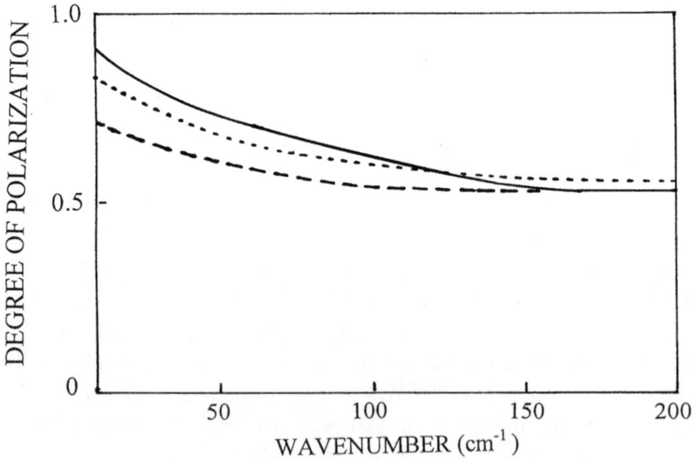

Figure 7 Degree of polarization of synchrotron radiation from the UVSOR ring. The solid line is the observed curve. Calculated lines are obtained with a vertical acceptance angle $2\psi = 60$ mrad (dashed curve) and with 40 mrad (dotted one)

We calculated the effective acceptance angle as a function of wavelength so that it reproduces the observed degree of polarization shown in Figure 7. The result is shown in Figure 8. The effective angle of acceptance at 10 cm^{-1} is less than half of the geometric one. This result indicates that the loss of light by diffraction is an important factor to be considered in the long-wavelength region.

3.3 SPECTROSCOPY UNDER HIGH PRESSURE

In the usual methods to obtain far-infrared spectra of samples under high pressure, a sapphire anvil cell is used. The highest pressure used in the experiments was limited to <0.5 GPa because of a lack of brilliant light sources. On the other hand, the SR source supplies a very brilliant light beam into a small solid angle, as shown in Figure 5. At the BL6A1 of UVSOR, high-pressure measurements using a diamond anvil cell (DAC) have been undertaken since 1989. Recently, a high pressure of up to 21 GPa was generated in an experiment on surface phonon absorption in microcrystals of various kinds of semiconductors. A schematic diagram of the optical configuration around the sample chamber for spectroscopy under high pressure and a cross-sectional view of the sample chamber including a lever-arm type of DAC combined with a liquid helium-cooled cryostat are shown in Figure 9. The DAC consists of a

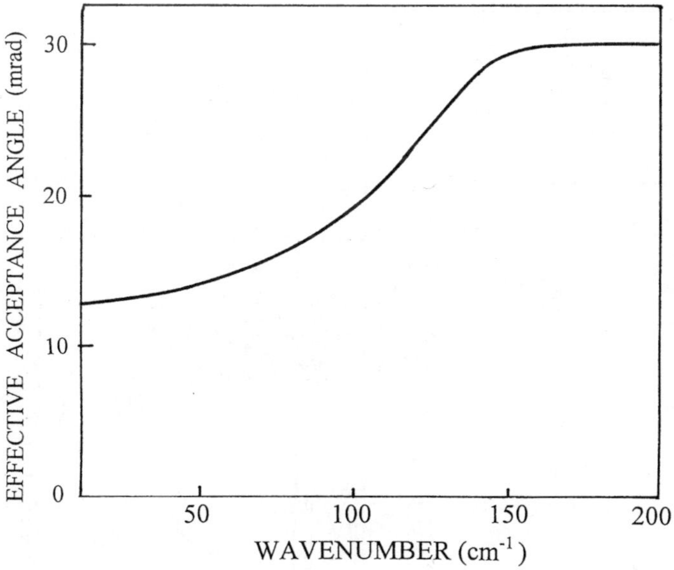

Figure 8 Effective acceptance angle (ψ) in the vertical direction to explain the observed degree of polarization of the synchrotron radiation given in Figure 7 (see text for details)

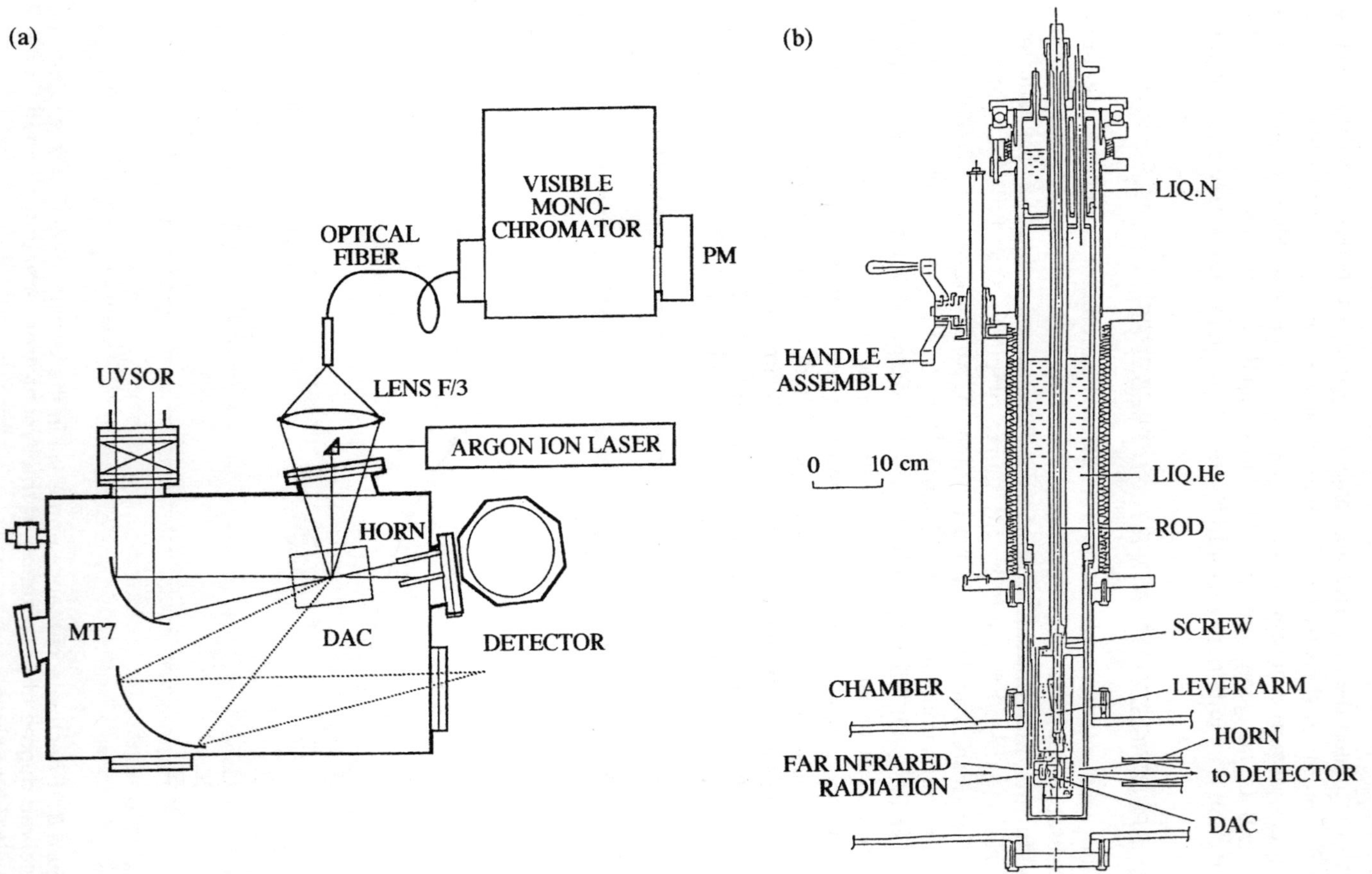
(a)
UVSOR
OPTICAL FIBER
VISIBLE MONO-CHROMATOR
PM
LENS F/3
ARGON ION LASER
HORN
MT7
DAC
DETECTOR
(b)
LIQ.N
HANDLE ASSEMBLY
0
10 cm
LIQ.He
ROD
SCREW
CHAMBER
LEVER ARM
HORN
FAR INFRARED RADIATION
to DETECTOR
DAC

pair of 0.25 carat diamonds whose flat face is 1.0 mm in diameter, anvil-mount plates made of tungsten carbide, a piston and alignment devices for the anvil. A stainless-steel plate was used as a gasket. The thickness and the diameter of the pinhole of the gasket of the DAC were 0.15 and 0.5 mm, respectively. In the DAC the sample was immersed in Apiezon-N grease, which works as a pressure-transmitting material. The grease was found to be almost transparent in the far-infrared region. As a sensor of the pressure generated in the DAC, ruby chips were inserted in the cell together with samples. The pressure generated in the cell was estimated by measuring of the energy shifts of the fluorescence R1 line of ruby which was excited by an argon ion laser. Recently, a reflection measurement under higher pressure using a DAC was developed at the BL6A1 of UVSOR. A light beam of UVSOR with a small source size and good divergence has been found to be suitable for an experiment under high pressure with a DAC; such a measurement is difficult to carry out with conventional light sources. The pressure monitor system is also shown schematically in Figure 9. In addition, the highly polarized light beam of UVSOR is also suitable for the spectroscopy under high pressure on specimens which show strong anisotropy along the crystal axis.

3.4 *EXTENSION TO MILLIMETER WAVE REGION*

In the optical configuration given in Section 3.1, the spectral intensity decreased steeply below 10 cm^{-1}. This is caused by Drude absorption by free carriers of a small amount of impurity contained in the silicon window. To increase the intensity in the long-wavelength region, the silicon window W1 in Figure 1 was replaced with crystalline quartz, which is transparent below 250 cm^{-1} and has a transmissivity of about 75% in this region. The wire grid beamsplitters were replaced with coarser ones with a 10 μm wire diameter and 500 μm spacing in order to limit the higher wavenumber components of the radiation and maintain the efficiency for the low-wavenumber region below 20 cm^{-1}. The spectrum was measured with an InSb hot electron detector, which has a higher sensitivity than a Ge bolometer in the low-wavenumber region. The intensity of the spectral distribution of the UVSOR below 30 cm^{-1} measured with the InSb detector is shown in Figure 10(a). The spectrum has a peak around 11 cm^{-1} and extends down to 3 cm^{-1}. The lower limit at 3 cm^{-1} is mainly due to the decrease of the sensitivity of the detector in addition to the

Figure 9 (a) Schematic diagram of the set-up around the sample chamber used for transmission measurements using the diamond anvil cell (DAC) and for pressure measurements with ruby fluorescence excited by an argon ion laser. (b) Cross-sectional view of the liquid helium cryostat containing the DAC and its support structure attached to the sample chamber maintained at 10^{-6} Pa

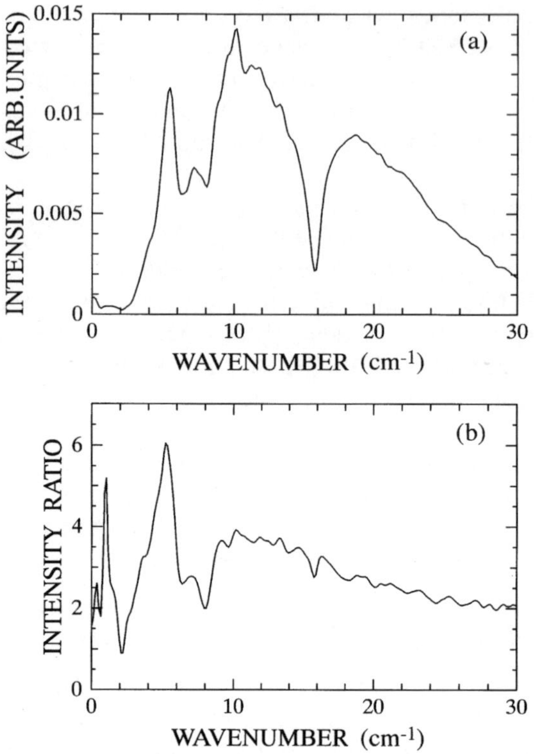

Figure 10 (a) Spectral intensity in the long-wavelength region including the millimeter wave region. The spectral resolution is 0.5 cm^{-1} and the detector is an InSb operated at liquid helium temperature. (b) Ratio of the intensity of UVSOR operating at 200 mA to that of a mercury lamp. The sharp dips at 6.5 and 8.0 cm^{-1} are due to the impurity absorption of the crystalline quartz window

diffraction loss, which is an important factor to be considered at long wavelengths. The large dips at 6.5 and 8.0 cm^{-1} are due to the impurity absorption of the crystalline quartz window. Figure 10(b) shows the ratio of the intensities in the spectral distribution of the UVSOR to that of a mercury lamp which is equipped with an interferometer using the same spectroscopic system without any aperture at the sample position. The intensity of the UVSOR was normalized to that at the ring current of 200 mA, which is the injection current in normal operation of the UVSOR. It can be seen that the intensity of the UVSOR is superior to that of the mercury lamp over the whole region although it is necessary to eliminate the impurity absorption so as to utilize the advantage of the UVSOR over the whole energy region.

No coherent enhancement of the intensity [16] or non-linear dependence of the intensity on the beam current [17] was observed in the millimeter wave region. This is due to the long electron bunch length of about 12 cm in the UVSOR ring.

4. APPLICATIONS

As indicated in the previous sections, the UVSOR was proved to be very useful in measurements on tiny specimens which show strong anisotropic properties owing to its high brilliance and high degree of polarization. We describe here some experiments as examples of the application of the UVSOR to interferometric measurements in order to demonstrate the advantages of its utilization over other sources.

4.1 PHONON SPECTRUM ON ALKALIMETAL–SILVER AND ALKALIMETAL–CUPROUS HALIDE CRYSTALS

Alkalimetal–silver and alkalimetal–cuprous halides crystals are compounds whose formula can be expressed as A_2MX_3 where A denotes an alkali metal ion, M a silver or a cuprous ion and X a halogen ion. The crystal structure was investigated by Brink and co-workers and found to have an orthorhombic structure with D_{2h}^{16} symmetry,containing chains of MX_4 tetrahedra along the *b*-axis as shown typically for K_2CuCl_3 in Figure 11 [18–20]. These crystals

Figure 11 Schematic diagram of the chain of tetrahedral $CuCl_4$ molecules along the *b*-axis of a K_2CuCl_3 crystal. Open large circles correspond to chloride ions and dashed circles to cuprous ions. A cuprous ion is surrounded by four chloride ions which form a slightly deformed tetrahedron connected to each other along the *b*-axis. Potassium ion, which is not shown, fills the interstitial spaces between the tetrahedra

contains four tetrahedra and 24 atoms per unit cell. According to factor group analysis on the lattice vibration mode of these crystals at $k=0$ (the Γ-point of the first Brillouin zone), the symmetries of the 72 vibration modes are resolved into the following modes with irreducible representation:

$$12A_g + 12B_{1g} + 6B_{2g} + 6B_{3g} + 6A_u + 6B_{1u} + 12B_{2u} + 12B_{3u}$$

Among these, three B_u modes are infrared active. Each mode contains one acoustic branch, and consequently the B_{1u} mode has five infrared-active phonon modes, and the others B_{2u} and B_{3u}, have 11 modes. We may resolve them in the spectrum if they are not degenerate in energy.

A few years ago only a needle-like crystal could be obtained, but now a single crystal with a size of a several mm^3 is obtainable for some compounds with careful sample preparation by slow cooling of a warm, saturated and stoichiometric aqueous solution. These crystals possess strong anisotropy in the optical properties along each crystal axis. For example, these crystals were found to show strong anisotropic properties in the optical spectrum in the exciton region [21, 22]. The optical spectrum has not been studied thoroughly because of the lack of a large single crystal, particularly the phonon spectrum. We carried out measurements of the anisotropic reflectivity along each crystal axis on these crystals by utilizing the high degree of polarization of the UVSOR.

Alkalimetal–silver halide crystals

We measured the anisotropic reflectivity of small single crystals of K_2AgI_3, Rb_2AgI_3, Rb_2AgBr_3 and Cs_2AgI_3 at 15 K. The effective diameter of the surface of the first three samples used for measurement was about 3 mm. For Cs_2AgI_3 we obtained only a needle-like thin single crystal and then measured the reflectivity in both polarization configurations, $E \parallel b$ (electric vector parallel to the b-axis) and $E \perp b$ (perpendicular to the b-axis). The spectrum of the $E \perp b$ configuration contains the $E \parallel a$ and $E \parallel c$ components. Other spectra were taken with linearly polarized light along the three crystal axes by rotating the direction of the output polarizer of the interferometer. The results are shown in Figure 12 for K_2AgI_3, Figure 13 for Rb_2AgI_3, Figure 14 for Rb_2AgBr_3 and Figure 15 for Cs_2AgI_3. For each sample, the dielectric constants, $\varepsilon = \varepsilon_1 + i\varepsilon_2$, were deduced from the Kramers–Kronig transformation of the reflection spectrum using the anisotropic high-frequency dielectric constants of each specimen reported by Edamatsu, Nanba and Ikezawa [22]. The real part (ε_1) and the imaginary part (ε_2) are also shown in Figures 12(b)–15(b), by solid and dashed curves, respectively.

Hitherto, a far-infrared transmission spectrum has been observed only for Cs_2AgI_3 in a series of this type of material by Bottger and Geddes [23]. The unpolarized transmission spectrum of a polycrystalline sample dispersed in

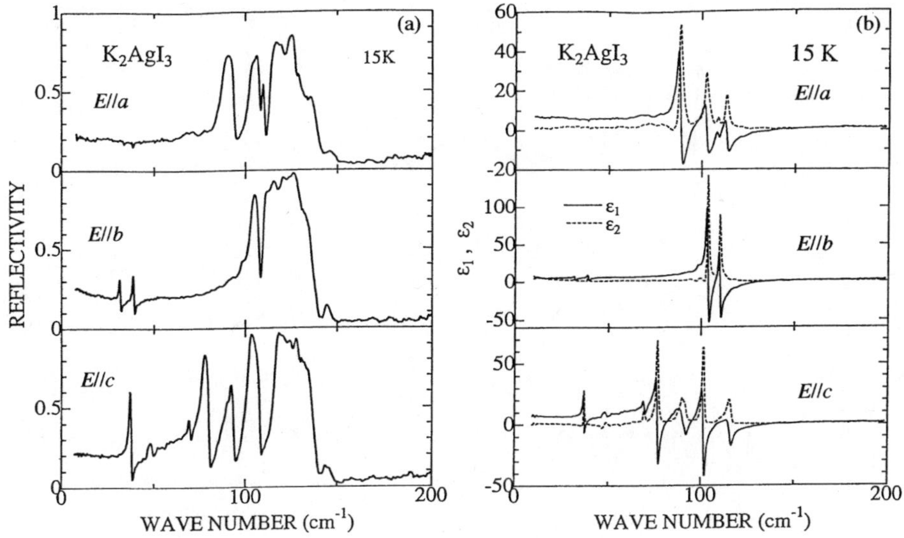

Figure 12 (a) Anisotropic reflection spectra of K_2AgI_3 crystal at 15 K for $E \| a$, $E \| b$ and $E \| c$. The spectral resolution was 1.5 cm^{-1}. (b) Real (ε_1) part (solid curves) and imaginary (ε_2) part (dashed curves) of the dielectric constants obtained through a Kramers–Kronig analysis of the reflection spectra

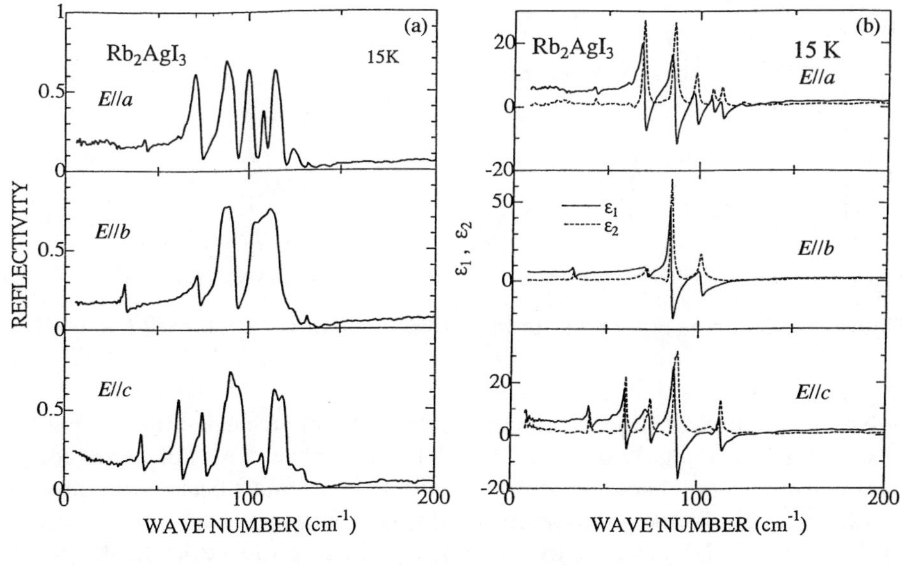

Figure 13 Same as Figure 12 but for Rb_2AgI_3 crystal

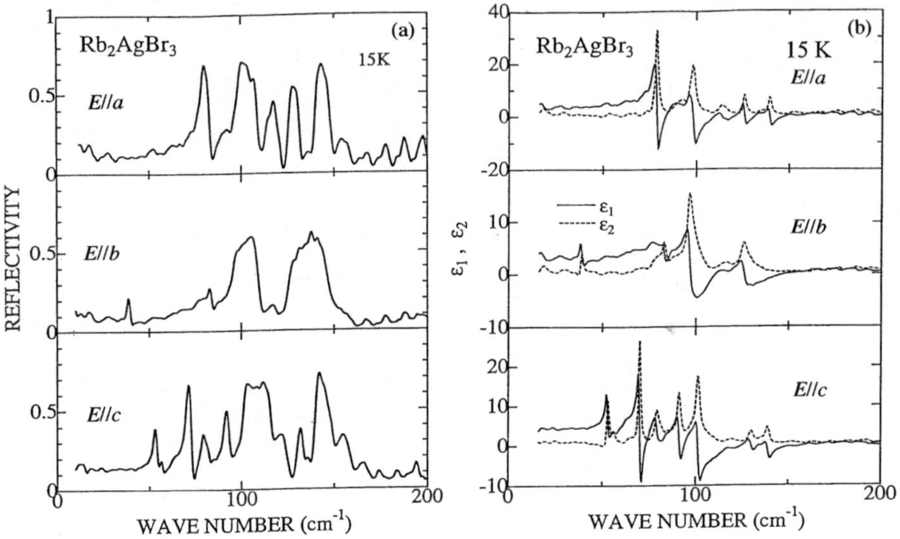

Figure 14 Same as Figure 12 but for Rb_2AgBr_3 crystal

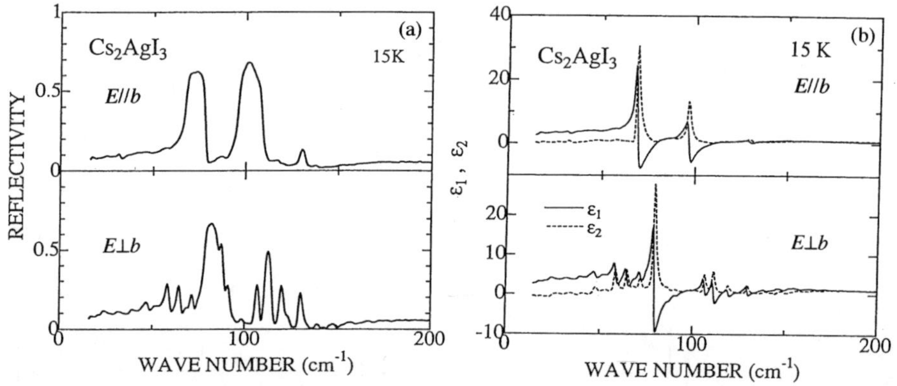

Figure 15 Same as Figure 12 but for Cs_2AgI_3 crystal at 15 K for $E \parallel b$ and $E \parallel b$

polyethylene discs at room temperature showed a multi-peak structure but the low energy band at 32 cm^{-1} in the $E \parallel b$ spectrum was not resolved. In our measurements, each spectrum exhibited a multi-peak structure as predicted in the factor group analysis. Replacement of the K^+ ion in K_2AgI_3 by Rb^+ (Rb_2AgI_3) gave an overall energy shift of the peaks to the lower energy side.

It is not easy to assign the resolved peaks in the reflection spectrum with a multi-peak structure to the symmetry of the phonon mode. The lowest energy band in each spectrum (e.g. the peaks at 30 cm^{-1} in the $E \parallel b$ spectrum and at 35 cm^{-1} in the $E \parallel c$ spectrum of K_2AgI_3, at 31 cm^{-1} in the $E \parallel b$ spectrum and at 40 cm^{-1} in the $E \parallel c$ spectrum of Rb_2AgI_3, at 40 cm^{-1} in the $E \parallel b$ spectrum and at 52 cm^{-1} in the $E \parallel c$ spectrum of Rb_2AgBr_3 and at 32 cm^{-1} in the $E \parallel b$ spectrum of Cs_2AgI_3) may be assigned to the intermolecular vibration between the MX_4 tetrahedra (shown in Figure 11) for following reasons. First, the energy positions of the these peaks are almost independent of kind of an alkali metal ion of the compounds and exist at nearly the same position in the corresponding polarization configuration among K_2AgI_3, Rb_2AgI_3, Rb_2AgBr_3 and Cs_2AgI_3. Second, the ratios of the positions of the lowest energy peaks, for example between Rb_2AgBr_3 and Rb_2AgI_3, are 1.27–1.30, which are almost the same as the expected value of 1.34 for the ratio of the intermolecular vibration energies between MX_4 tetrahedra which is given by

$$\omega(\mathrm{AgBr_3})/\omega(\mathrm{AgI_3}) = \sqrt{\mu(\mathrm{AgI_3})/\mu(\mathrm{AgBr_3})}$$

where μ is a reduced mass of the tetrahedron, from a simple lattice vibration calculation under the assumption that the force constant does not change.

Alkalimetal–cuprous halide crystals

Alkalimetal–cuprous halide crystals have an orthorhombic structure like alkalimetal–silver halide crystals. We measured the reflectivity of small single crystals of K_2CuBr_3, Rb_2CuI_3 and K_2CuCl_3 at 15 K. Spectra were taken with linearly polarized light along the three crystal axes by rotating the direction of the output polarizer of the interferometer. The results are shown in Figure 16 for K_2CuBr_3, Figure 17 for Rb_2CuI_3 and Figure 18 for K_2CuCl_3. For K_2CuCl_3, a single crystal was obtainable only in a needle-like form and then only the spectra in the polarization configurations $E \parallel b$ and $E \perp b$ were measured. Using a similar procedure to that for alkalimetal–silver halide crystals, the dielectric constants for each sample were deduced from Kramers–Kronig analysis of the reflection spectrum using the anisotropic high-frequency dielectric constants of each specimen reported by Edamatsu, Nanba and Ikezawa [21]. The real part (ε_1) and the imaginary part (ε_2) are also shown by solid and dashed curves in Figures 16(b)–18(b), respectively. Each spectrum exhibited a multi-peak structure similar to those for alkalimetal–silver halide crystals. Replacement of Br^- ion in K_2CuBr_3 by the lighter Cl^- ion (K_2CuCl_3) gave an overall and clear energy shift of the main peaks from 120–170 to 150–200 cm^{-1} for the $E \parallel b$ and from 100–170 cm^{-1} for the $E \parallel a$ and $E \parallel c$ spectra to 150–230 cm^{-1} for the $E \perp b$ spectrum. This suggests that the majority of the main peaks of these crystals and alkalimetal–silver halide crystals originate from the intramolecular vibration modes within the MX_4 tetrahedron.

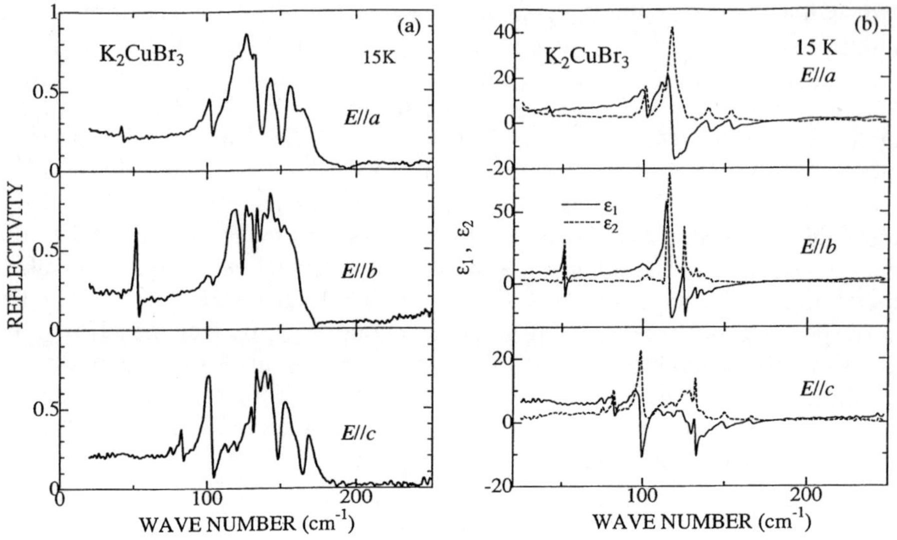

Figure 16 (a) Anisotropic reflection spectra of K_2CuBr_3 crystal at 15 K for $E \| a$, $E \| b$ and $E \| c$. The spectral resolution was 1.5 cm^{-1}. (b) Real (ε_1) part (solid curves) and imaginary (ε_2) part (dashed curves) of the dielectric constants obtained through a Kramers–Kronig analysis of the reflection spectra

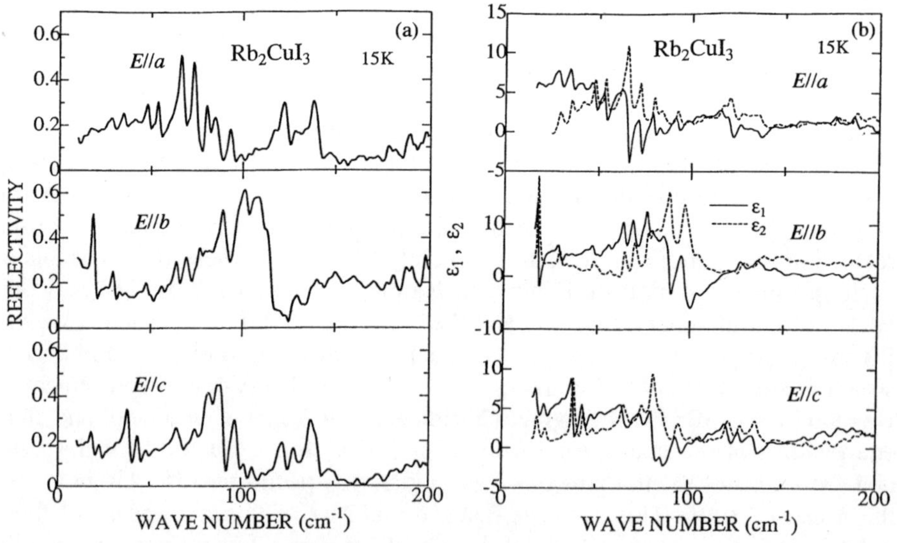

Figure 17 As Figure 16 but for Rb_2CuI_3 crystal

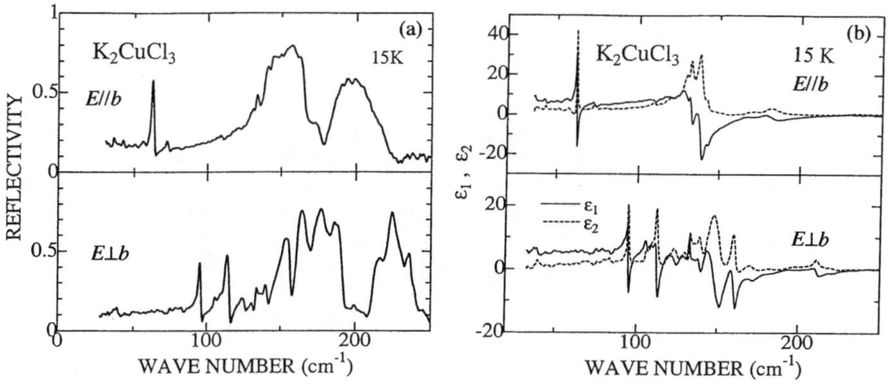

Figure 18 As Figure 16 but for K_2CuCl_3 crystal for $E \| b$ and $E \perp b$

4.2 *BLACK PHOSPHORUS UNDER HIGH PRESSURE*

Anisotropic phonon spectrum due to the B_{1u} mode

Black phosphorus (black P) is the most stable form of many allotropic modifications of phosphorus under normal conditions and is a narrow-gap semiconductor with a gap energy of about 0.3 eV [24]. Black P is composed of puckered layers and has an orthorhombic crystal structure (D_{2h}^{18}). The structure of the puckered layers is shown schematically in Figure 19(a) along with the Cartesian axes system. Black P shows strong two-dimensional properties along the $x-y$ plane in various physical phenomena owing to its strong anisotropy in its layer-like crystal structure. According to its crystal symmetry, two infrared-active phonon modes, B_{1u} and B_{3u} are predicted to exist at the Γ-point of the Brillouin zone [25]. Black P is monoatomic material similar to silicon and germanium and has no permanent dipole moment because the four atoms in the primitive cell are the same. However, an effective dipole moment can be induced by atomic displacement according to each infrared-active mode. The induced dipole moment due to the B_{1u} phonon mode is shown schematically in Figure 19(b). The anisotropic absorption on the B_{3u} phonon mode was observed at 470 cm^{-1} by a polarized infrared absorption measurement on a single thin crystal and was found to possess an induced dipole moment in the direction along the z-axis as predicted theoretically [26]. In addition, the existence of a small energy splitting between the TO and LO phonon energies was found [27].

Quantitative measurements on the B_{1u} mode have not been performed until now because of the lack of a single crystal large enough for far-infrared

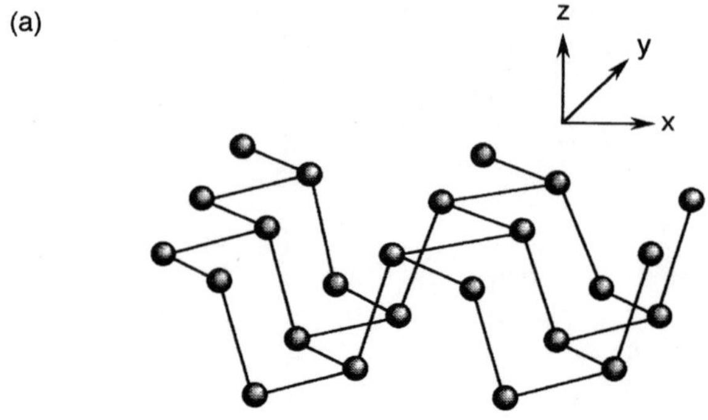

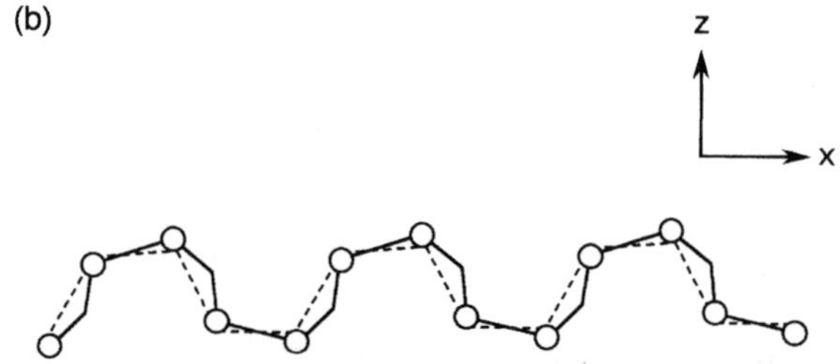

Figure 19 (a) Structure of a single puckered layer of black phosphorus along with the Cartesian coordinate system. (b) Schematic diagram of the atomic displacement according to the B_{1u} mode projected on to the x–z plane. The open circle corresponds to the phosphorus atom and the closed circle to the induced charge due to the atomic displacement [25]. See text for details

measurements. It was difficult to obtain a large and homogeneous single crystal because of the crucial conditions for the crystal growth method under high pressures of 1–2 GPa at high temperature. An intense band which is assigned to the B_{1u} phonon mode was found to exist at 127–136 cm^{-1} in the foregoing measurements of the absorption spectrum [26, 28] and the reflection spectrum [27]. Although a small energy splitting between the TO and LO phonon energies was found to exist, further quantitative analysis was not carried out because of the difficulty of precise measurements on a small specimen using a conventional light source in the far-infrared region. We used SR at the UVSOR facilities as the light source and carried out anisotropic reflection measurements on a phosphorus single crystal at low temperature to obtain a high-quality spectrum for the B_{1u} phonon mode.

A single crystal of black P was cleaved in dried air. The cleaved surface showed a dark metallic luster. The x- and y-axes in the cleaved surface were determined by the X-ray diffraction method. The effective area of the surface used for the measurement was about 3 mm in diameter. The reflection spectra of the single crystal were measured at 15, 80 and 300 with linearly polarized light whose electric vector was parallel to the cleaved $x-y$ surface of the crystal. A strong anisotropic reflection spectrum obtained at 15 K is shown in Figure 20. The solid curve was measured with linearly polarized light whose electric vector was parallel to the x-axis of the crystal. This configuration is denoted $E \parallel x$. A large peak, whose form is similar to that observable in a typical ionic crystal, was resolved at around 132.0 cm^{-1} and an additional peak at 140.0 cm^{-1}. A large bandwidth in the reflection band suggests the existence of a Lyddane–Sachs–Teller (LO–TO) splitting of the phonon mode. The dashed curve was measured at 15 K in the $E \parallel y$ configuration by rotating the direction of the output polarizer of the interferometer by 90°, and no distinct peak was resolved. These results confirmed the theoretical prediction that the dipole moment according to the atomic displacement due to the B_{1u} mode is induced in the direction of the x-axis of the crystal.

The dielectric constant, $\varepsilon = \varepsilon_1 + i\varepsilon_2$, at 15 K was deduced from the Kramers–Kronig transformation of the $E \parallel x$ reflection spectrum. The resulting imaginary part (ε_2) of the dielectric constant and the energy loss function [$Im(1/\varepsilon)$] are shown in Figure 21(a) and (b). The peak position in the spectrum of the loss function corresponds to the LO phonon energy. From the peak position of the imaginary part (ε_2), the TO phonon energy ω_{TO} was determined to be 127.5 cm^{-1}. From the peak position of the energy loss

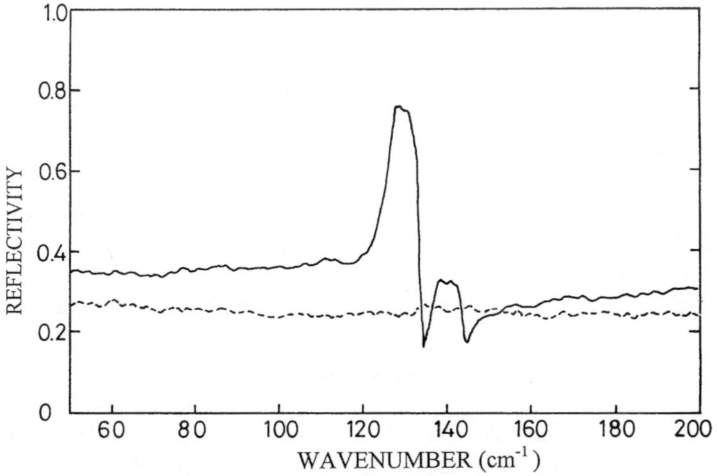

Figure 20 Reflection spectra of a black phosphorus single crystal at 15 K. The effective area of the sample surface used for the measurements was about 3 mm in diameter. The solid curve was measured in the $E \parallel x$ configuration and the dashed curve in the $E \parallel y$ configuration. The spectral resolution was 0.64 cm^{-1}

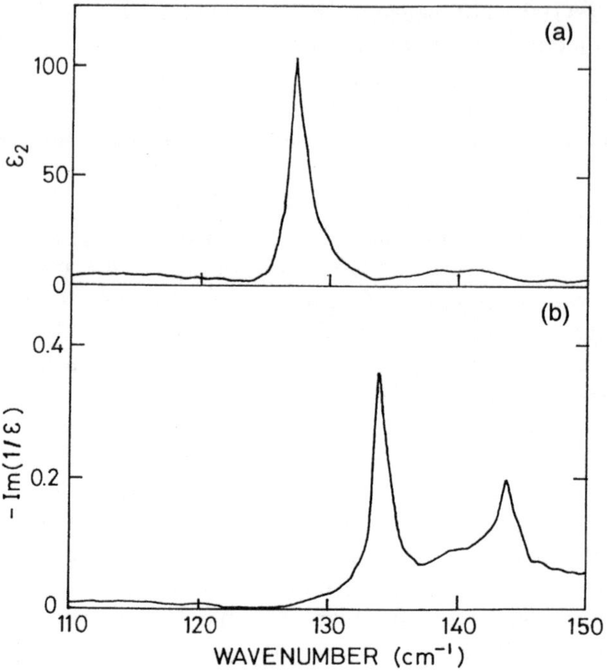

Figure 21 (a) Imaginary (ε_2) part of the dielectric constants and (b) the energy loss function [$Im(1/\varepsilon)$] of a black phosphorus single crystal obtained through a Kramers–Kronig analysis of the reflection spectra for $E \| x$ in Figure 20

function, the LO phonon energy ω_{LO} was determined to be 133.9 cm^{-1} . The amount of Lyddane–Sachs–Teller (LO–TO) splitting, which is given by $\Delta = \omega_{LO} - \omega_{TO}$, is estimated to be 6.4 cm^{-1} at 15 K. The values of ω_{LO}, ω_{TO} and Δ obtained are in an excellent agreement with the theoretically calculated values [25]. Thus, the phonon energies of black P were determined precisely by the present experiment. Black P is probably the first material for which the Lyddane–Sachs–Teller splitting was observed in a homogeneous covalent crystal.

The measurement of an anisotropic static dielectric constant (ε_0) of black P has not yet been reported. We obtained it by extrapolating the low-energy part of the observed reflectivity to the zero-frequency limit (R_0) under the assumption that no other dispersion exists in the energy below the 'reststrahlen' band. We obtain

$$\varepsilon_0 = (1 + \sqrt{R_0})^2/(1 - \sqrt{R_0})^2$$

The values obtained are $\varepsilon_0(E \| x) = 14.8$ and $\varepsilon_0(E \| y) = 10.6$, and are in good agreement with those obtained theoretically [29].

In black P, the static effective charge is zero because the four atoms in the primitive unit cell are the same. However, the macroscopic effective charge (e_T^*) is induced by the atomic displacement due to the B_{1u} phonon mode, which is given by

$$(e_T^*)^2 = \varepsilon_\infty(\omega_{LO}^2 - \omega_{TO}^2)/(4\pi Ne/M)$$

where e_T^* is the dimensionless macroscopic effective charge [30], e the electron charge, N the atomic density in the unit cell, M the mass of a phosphorus atom and ε_∞ the high-frequency dielectric constant. In black P, $N = 5.214$ cm^{-3} from the values of the lattice constants [31] and $M = 5.19 \times 10^{-23}$ g. Using these values and $\varepsilon_\infty = 10$ which is given by the calculation, e_T^* was estimated to be 0.44 at 15 K for the B_{1u} phonon mode. The magnitude is about one third of those the typical ionic crystals such as alkalimetal halide crystals.

Temperature dependence of the reflection spectrum

The reflection spectrum due to the B_{1u} phonon mode was also measured at 80 and 300 K. The spectrum at 300 K is shown in Figure 22 by the dashed curve in comparison with the solid curve obtained at 15 K. The peak showed a clear energy shift to the lower energy side ('red shift') on cooling of the specimen to 15 K. The values of ω_{TO}, ω_{LO}, Δ and e_T^* at 80 and 300 K were obtained from the dielectric constants, which were deduced from the Kramers–Kronig analysis of the reflection spectra at 80 and 300 K by the same procedure as that

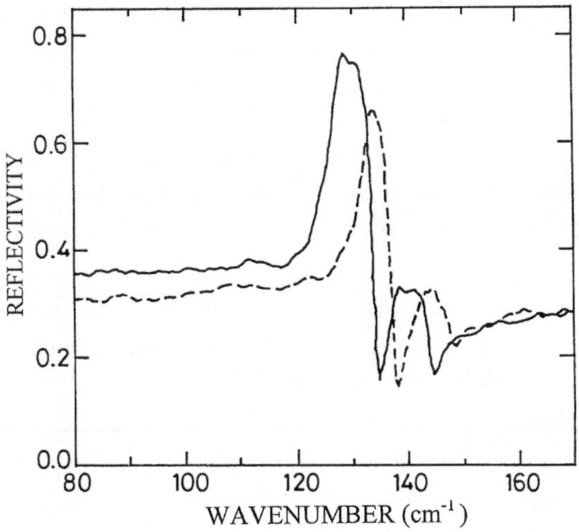

Figure 22 Reflection spectra of a black phosphorus single crystal due to the B_{1u} mode. The solid curve was measured at 15 K and the dashed curve at 300 K

used at 15 K. At 80 K the values obtained are $\omega_{TO} = 128.7\ cm^{-1}$, $\omega_{LO} = 134.1\ cm^{-1}$, $\Delta = 5.4\ cm^{-1}$ and $e_T^* = 0.42$ and at 300 K they are $\omega_{TO} = 133.2\ cm^{-1}$, $\omega_{LO} = 137.3\ cm^{-1}$, $\Delta = 4.1\ cm^{-1}$ and $e_T^* = 0.37$. The values of the TO and LO phonon energies, the Lyddane–Sachs–Teller splitting (Δ) and the macroscopic effective charge (e_T^*)are summarized in Table 1.

The energies of the TO and LO phonons are plotted against temperatures in Figure 23. The TO phonon energies in the foregoing experiments at different temperatures are different from each other [26–28] and the energy differences are beyond the experimental error. It was found in the present experiments that such a disagreement in the TO phonon energy in the foregoing experiments was caused by the anomalous peak shift with temperature. In black P, the peak showed a clear red shift which cannot be interpreted by the usual theory on anharmonicity of lattice vibrations. A peak shift and a line broadening of a phonon band with temperature are usually interpreted by anharmonicity in the lattice vibration. A decrease in an anharmonicity by cooling of specimens will nominally give rise to the blue shift.

The static dielectric constants were also obtained at 80 and 300 K using a procedure similar to that used at 15 K. The values are $\varepsilon_0(E \parallel x) = 14.4$ at 80 K and $\varepsilon_0(E \parallel x) = 13.0$ and $\varepsilon_0(E \parallel y) = =10.6$ at 300 K. The values of the static dielectric constants obtained in the present experiments are summarized in Table 2.

In the present measurements a secondary peak was observed at the higher energy side of the main structure in the $E \parallel x$ configuration . The peak positions in the ε_2 spectrum of the dielectric constant and in the energy loss function $[Im(1/\varepsilon)]$ with temperature are also plotted in Figure 23 as squares. The peak positions showed a strong temperature dependence similar to those in the main structure. Black P possesses no other infrared-active phonon mode in the first-order phonon process than the B_{1u} and B_{3u} modes. In addition, these additional peaks cannot be attributed to a two-phonon band because the intensity ratios between the main and secondary peaks were nearly the same at each temperature. In general, a two-phonon band should grow with increase in temperature according to the Boltzmann factor in the phonon occupation number. For black P, the anomalous peak shifts with temperature and the origin of the secondary peaks are still a puzzle.

Table 1 TO and LO phonon energies, Lyddane–Sachs–Teller splitting (Δ), and macroscopic effective charge (e_T^*) of black phosphorus at 300, 80 and 15 K due to the B_{1u} phonon mode

Temperature (K)	TO (cm^{-1})	LO (cm^{-1})	Δ (cm^{-1})	e_T^*
300	133.2	137.3	4.1	0.37
80	128.7	134.1	5.4	0.42
15	127.5	133.9	6.4	0.44

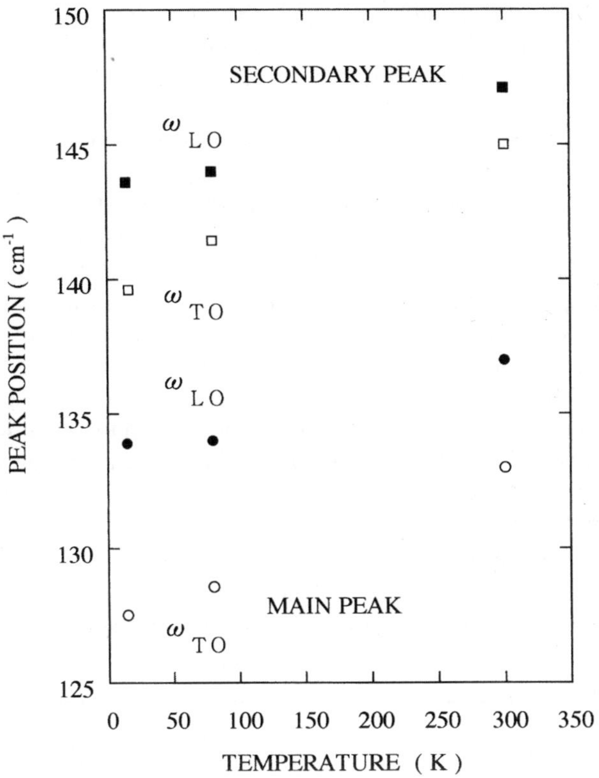

Figure 23 Plot of the energies of the TO and LO phonons due to the B_{1u} mode as a function of temperature for the main peak (circles) and for the secondary peak (squares). Open symbols correspond to the TO phonon mode and closed symbols to the LO phonon mode

Table 2 Anisotropic static dielectric constants (ε_0) of black phosphorus

Temperature (K)	$\varepsilon_0(E \parallel x)$	$\varepsilon_0(E \parallel y)$
300	13.0	10.6
80	14.4	—
15	14.8	10.6

Metal–insulator transition under pressure

Black P undergoes a change in electronic state under pressure. Under normal conditions, black P is a semiconductor and shows a clear reststrahlen band in the phonon reflection spectrum, as shown in Figure 27. According to the measurement of electric conductivity [32, 33], black P shows a metal–insulator transition at around 4 GPa at a room temperature. A region of 'semimetallic phase' exists in the pressure range 4–10 GPa. Above 10 GPa, black P shows almost metallic properties and finally shows a superconducting transition at about 17 GPa. The change in the crystal structure under pressure was measured by X-ray diffraction [34].

We measured the anisotropic transmission spectrum of the B_{1u} phonon mode of black P at room temperature. Figure 24 shows the spectra at normal pressure (P = 0) and at 1.2 GPa. At $P = 0$, a sharp dip in the absorption due to the B_{1u} phonon mode appeared at 133 cm^{-1}. A small periodic structure superposed on the spectrum was due to the interference fringes of the cleaved thin sample. At $P = 1.2$ GPa the distinct peak disappeared. This is considered to be due to the insulator–metal transition of black P under pressure. The remaining intensity in the spectrum was due to the leakage of the incident light from the small blank space in the pressure cell. In order to follow the spectral change with pressure, the spectra were measured under different pressures as shown in Figure 25. The distinct dip structure at 133 cm^{-1} at $P = 0$ showed a clear red shift and broadening with pressure. A red shift with pressure is similarly seen in the spectrum with cooling of the specimen shown in Figure 22. The pressure played a same role in the peak shifts of the B_{1u} phonon band of black P with

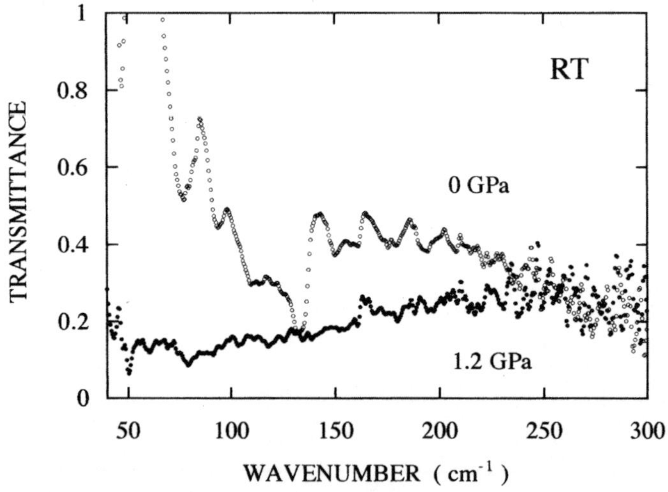

Figure 24 Transmission spectra of a black phosphorus single crystal at room temperature due to the B_{1u} mode for the $E \| x$ configuration at normal pressure ($P = 0$) and at 1.2 GPa. The spectral resolution is 1.6 cm^{-1}. The remaining intensity in the spectrum at 1.2 GPa corresponds to the leakage of the light beam from the blank space which metallic black P does not cover in the pressure cell

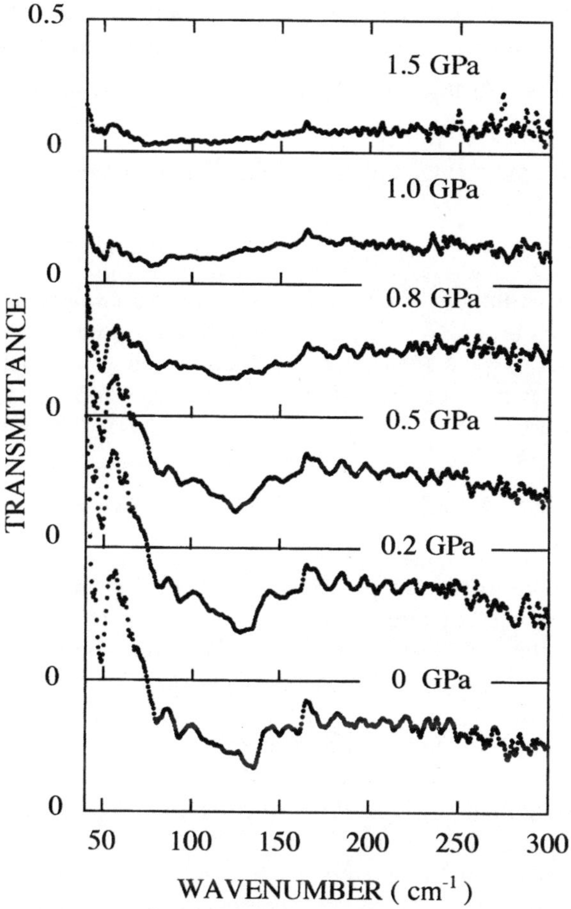

Figure 25 Development of transmission spectra of a black phosphorus single crystal at room temperature due to the B_{1u} mode with pressure. Applied pressure is shown in the figure. The spectra were recorded with a resolution of 1.6 cm^{-1}

temperature. At $P = 1.0$ GPa the distinct peak disappeared and a structureless continuum remained. This means that black P changed from an insulator to a semi-metal phase at $P = 1.0$ GPa. According to the literature [32, 33], a semi-metal phase was found to exist at around 4–10 GPa at room temperature. However, the present experiments showed that the phase change in black P to a semi-metal phase occurred at a lower pressure than that observed in the electric resistivity measurements. This suggests that optical measurement is very sensitive to a change in an electronic state and is a useful method to detect a precursor phenomenon in a phase change. The mode Gruneisen constant of the

B_{1u} phonon mode was deduced from the peak energy shifts with pressure. The γ_i value of a phonon mode i is defined by the following equation:

$$\gamma_i = -(\mathrm{d}\log\omega_i/\mathrm{d}\log V) = -(\mathrm{d}\omega_i/\omega_{0i})/(\mathrm{d}V/V_0)$$

where $\mathrm{d}\omega_i$ is the energy shift of the phonon, ω_{0i} is the phonon energy at $P=0$ and $\mathrm{d}V$ is the volume change under pressure. The volume change with pressure is given by

$$\mathrm{d}V = V - V_0 = -\beta V_0\,\mathrm{d}P$$

where V_0 is the volume at $P=0$ and β is a volume compressibility. We take $\beta = 0.028$ GPa^{-1} [34]. At $P=0.5$ GPa, the peak appeared at 123.4 cm^{-1}. Then we obtained $\gamma(B_{1u}) = -5.16$, which is negative and is in contrast to normal ionic insulators because of the red shift of the absorption of the B_{1u} phonon mode with pressure.

4.3 MILLIMETER WAVE ABSORPTION OF GERMANIUM CRYSTAL

The cross-sectional diameter of the light beam for the measurement was focused to 3 mm with a large F-number optics of $F/6$ in the far-infrared region, as described in Section 3.1. Therefore, the spectroscopic system at UVSOR is very convenient for measurements on small specimens. Here, we give an example of measurements in the millimeter wave region to show the usefulness of SR also in the long-wavelength region.

A pure germanium crystal is known to show absorption at room temperature in the low-wavenumber region below 30 cm^{-1} [35]. This absorption has been interpreted as being due to the intrinsic free carriers which are thermally excited. Using wire grid polarizers for the long-wavelength region and the crystal quartz window, absorption spectra of a germanium crystal were measured at 300, 80 and 13 K. To obtain a net absorption, signal intensities were observed on two samples with different thicknesses of $d_1 = 2.7$ and $d_2 = 0.6$ mm which were cut to a wedge shape. The net absorption coefficient is given by $\alpha = \ln(\Delta I)/\Delta d$, where $\Delta d = d_1 - d_2$ is the difference in the sample thickness and $\Delta I = I_2(d_2) - I_1(d_1)$. The result is shown in Figure 26. At 300 K a distinct peak was observed at around 6 cm^{-1} and decreased in intensity at 80 K, which indicates that the absorption is not due to the free carriers. The absorption almost disappeared at 13 K. The temperature dependence suggests that it can be interpreted as an absorption band due to a multi-phonon difference process. In general, a phonon dispersion curve obtained by a neutron diffraction method is useful for assigning an observed peak to a special phonon branch. However, in the present case it is difficult to assign definitely the origin of the absorption band at 6 cm^{-1} because of the lack of the accuracy in the energy resolution of the neutron experiment on measurement of the phonon dispersion curve [36].

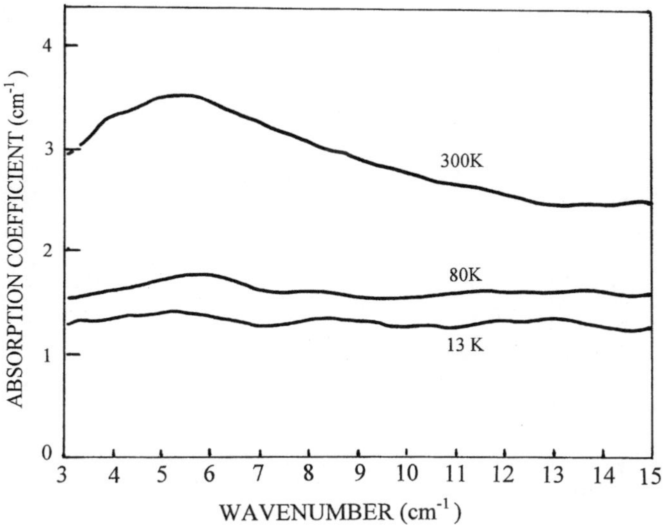

Figure 26 Absorption spectra of Ge crystal in the long-wavelength region at 300, 80 and 13 K

5 FURTHER PERSPECTIVES

As described in Section 3, the observed beam quality of SR emitted from the bending dipole magnet of the storage ring has a high degree of polarization. Here, we consider conditions for achieving SR with high polarization and more intensity at a storage ring.

Figure 27(a) shows the computed spectral distribution curves of SR as a function of different acceleration energies E of electrons with the same bending radius ($\rho = 3.3$ m). It can be seen that the intensity of the far-infrared part does not change with the acceleration energy. An acceleration energy controls the lifetime of running electrons in the ring and is essentially related to a long lifetime. Computed spectral distribution curves as a function of different bending radii (ρ) with the same acceleration energy $E = 1.5$ GeV are shown in Figure 27(b). The calculated curve shows that the emitted photon numbers per unit current (1 mA) grows with the factor in proportion to $\rho^{1/3}$ in the longer wavelength region. Although the intensity decreases gradually with increase in wavelength according to the factor of $\lambda^{-4/3}$ for each curve, it is independent of the acceleration energy, as seen in the theoretical prediction [4]. The computed beam spread of the SR with a wavelength of 100 μm emitted at a vertical angle (ψ) to the ring with a unit horizontal angle ($\theta = 1$ mrad) is shown as function of ρ in Figure 27(c). This figure clearly shows that more intense SR from a magnet with a larger bending radius of the

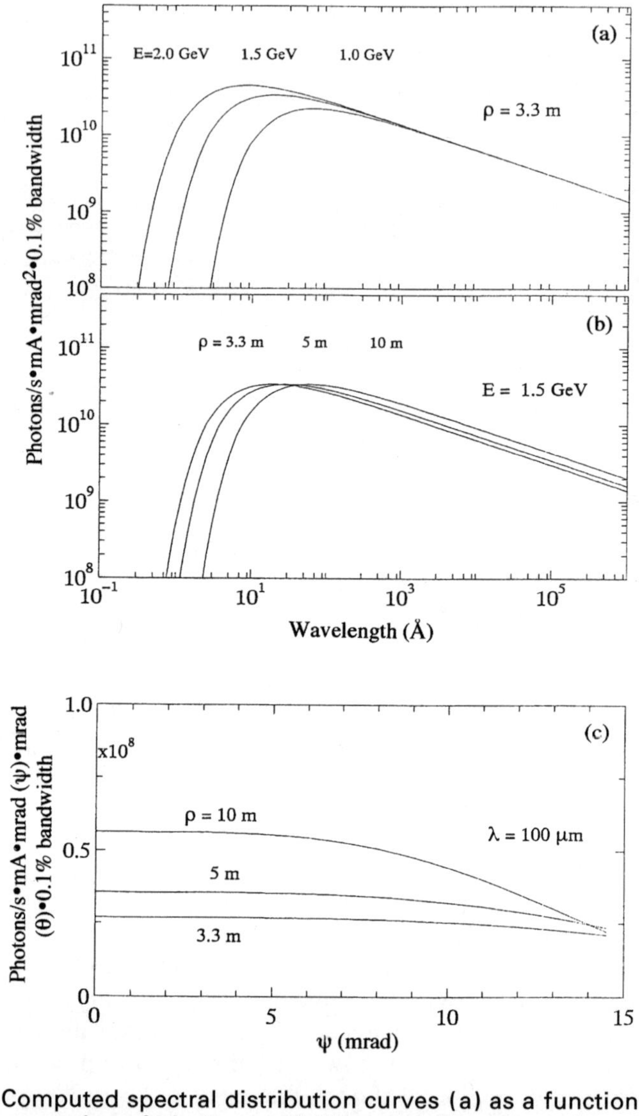

Figure 27 Computed spectral distribution curves (a) as a function of different acceleration energies of electron with the same bending radius of 3.3 m and (b) as a function of different bending radii with the same acceleration energy of 1.5 GeV and (c) computed vertical angle distribution curve of photon numbers at 100 μm as a function of different bending radii with the same parameters as in (b)

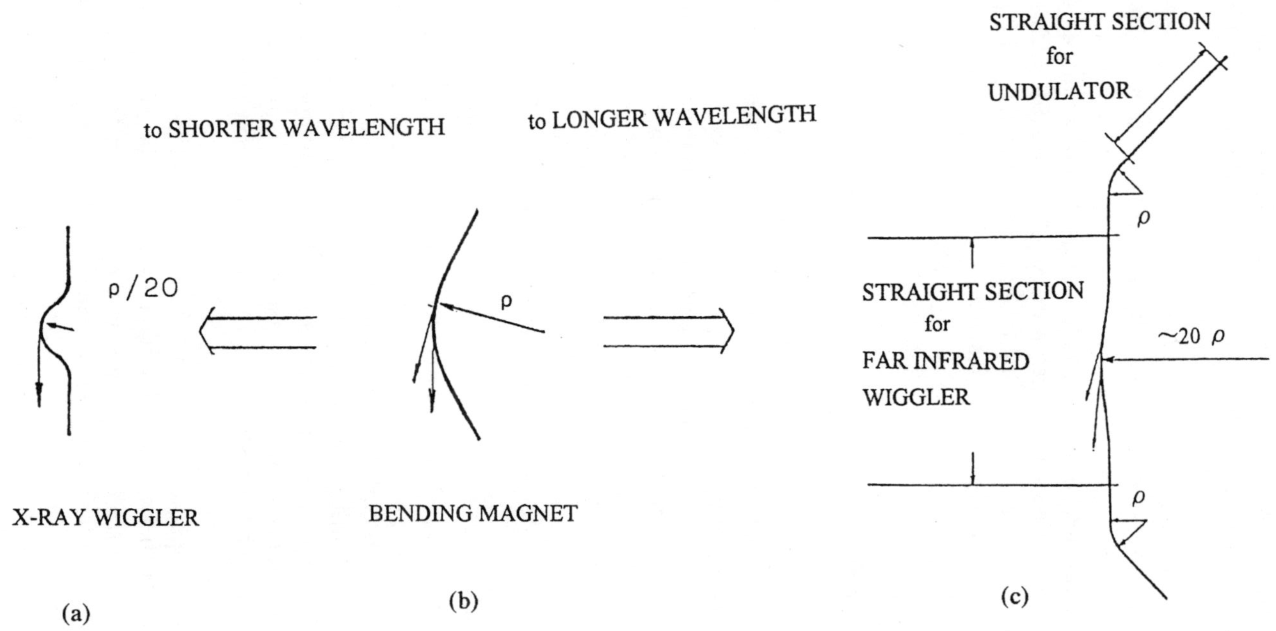

Figure 28 Schematic diagram of a single period wiggler as a wavelength shifter installed in the straight section of a storage ring. (a) An X-ray wiggler as a shorter wavelength shifter; (b) a normal bending section; (c) a far-infrared wiggler as a longer wavelength shifter

orbit of electrons is emitted into a narrower vertical angle. The intensity of the parallel component of the light beam concentrates closer to the $\psi = 0$ plane and its relative intensity grows more than the perpendicular component. This suggests that the polarization degree and the brilliance of the emitted beam increase with increase in ρ. Such an effect is favorable for interferometric measurements with SR.

One of the easiest ways to realize such a highly-brilliant and highly polarized far-infrared light beam is to install a single-period wiggler with a larger bending radius of a few tens of meters in the straight section of the storage ring as a wavelength shifter towards a longer wavelength region [9]. Such a device installed in the section of the storage ring is shown schematically in Figure 28. Hitherto a common X-ray wiggler was installed as a wavelength shifter towards a shorter wavelength region.

6 CONCLUSIONS

Synchrotron radiation has been demonstrated to be a very powerful and useful source for far-infrared and millimeter wave spectroscopy. There is no other source in these regions which has a completely continuous and structureless spectrum like SR except for BB radiation, which is, however, inferior to SR radiation in both intensity and brightness. The cross-section of the light beam obtained at the UVSOR facility is about 3 mm in the far-infrared region. It can be reduced to <1 mm by using an additional focusing device. Experiments in a small cell such as a diamond anvil cell for spectroscopy under high pressure are possible. Far-infrared spectroscopy under high pressures up to 21 GPa has been performed at the UVSOR facility.

The above property is due to the high brightness of SR. The polarized property is also convenient for a spectrometer such as the Martin–Puplett type which utilizes polarized light. The high degree of polarization is also favorable for interferometric measurements on materials which show strong anisotropy in the optical properties. Further, the pulsed property will find many applications because there is no other source of far-infrared and millimeter wave light which is completely tunable, with such a short duration (sub-nanoseconds). Hence, we may conclude that SR is one of the most useful sources for far-infrared and millimeter wave spectroscopy.

The above useful properties all belong to radiation from the bending section of a storage ring which has been designed solely for the short-wavelength region without any consideration of use in long-wavelength regions. If we construct a device particularly designed for long-wavelength regions, e.g. a far-infrared wiggler with a suitably large bending radius, we will be able to obtain an improved source with higher brightness and intensity and high degree of polarization.

7 ACKNOWLEDGEMENTS

The work described in this chapter was carried out has been done in cooperation with the staff of UVSOR and we are grateful to all staff members for their kind collaboration in the experiments. The authors are grateful to Professor I. Shirotani of Muroran Institute of Technology for supplying high-quality black phosphorus single crystals. They are grateful also to Professor T. Awano of Tohoku Gakuin University for his support in the experiment on alkalimetal–silver and –cuprous halide crystals. They are grateful to Professor M. Watanabe of Tohoku University, who was on the staff of UVSOR at the beginning of the present work, and are also sincerely grateful to Professor H. Inokuchi, former Director of UVSOR, for his guidance and encouragement in the project on far-infrared spectroscopy at UVSOR.

8 REFERENCES

1. G.W. Chantry, *Long Wave Optics*, Vols 1 and 2, Academic Press, New York (1984).
2. D.H. Martin (Ed.), *Spectroscopic Technique*, North-Holland, Amsterdam (1967).
3. K.J. Button (Ed.), *Infrared and Millimeter Waves*, Vols 1–14, Academic Press, New York (1979–85).
4. J. Schwinger, *Phys. Rev.*, **75** 1912 (1949).
5. H. Winick and S. Doniac (Eds), *Synchrotron Radiation Research*, p. 27, Plenum Press, New York (1980).
6. E.-E. Koch (Ed.), *Handbook on Synchrotron Radiation*, p. 1, North-Holland, Amsterdam (1983).
7. M. Meyer and P. Lagarde, *J. Phys.*, **37**, 1387 (1976).
8. P. Lagarde, *Infrared Phys.*, **18**, 395 (1978).
9. J. Yarwood, T. Shuttleworth, J.B. Hasted and T. Nanba, *Nature (London)*, **312**, 743 (1984).
10. T. Nanba, J. Yarwood, T. Shuttleworth and J.B. Hasted, *Int. J. Infrared Millimeter Waves*, **7**, 759 (1986).
11. E. Schweizer, J. Nagel, W. Braun, E. Lippert and A.M. Bradshaw, *Nucl. Instrum. Methods*, **A239**, 630 (1985).
12. T. Nanba, Y. Urashima, M. Ikezawa, M. Watanabe, E. Nakamura, F. Fukui and H. Inokuchi, *Int. J. Infrared Millimeter Waves*, **7**, 769 (1986).
13. G.P. Williams, E.J. Hirschmugl, E.M. Kneedler, P.Z. Tacks, M. Shleifer, Y.J. Chabal and F.M. Hoffman, *Phys. Rev. Lett.*, **62**, 261 (1989).
14. J.R. Stevenson, H. Ellis and R. Barloett, *Appl. Opt.*, **12**, 2884 (1973).
15. T. Nanba, *Rev. Sci. Instrum.*, **60**, 1680 (1989).
16. F.C. Michel, *Phys. Rev. Lett.*, **48**, 580 (1982).
17. T. Nakazato, M. Oyamada, N. Nimura, S. Urasawa, O. Konno, A. Kagaya, R. Kato, T. Kamiyama, Y. Torizuka, T. Nanba, Y. Kondo, Y. Shibata, K. Ishi, T. Ohsaka and M. Ikezawa, *Phys. Rev. Lett.*, **63**, 1245 (1989).
18. C. Brink and C.H. MacGillavry, *Acta Crystallogr.*, **2**, 158 (1949).
19. C. Brink and H.A.S. Kroes, *Acta Crystallogr.*, **5**, 433 (1952).
20. M.M. Thackerey and J. Cestzer, *Acta Crystallogr.*, **31**, 2339 (1975).

21. K. Edamatsu, T. Nanba and M. Ikezawa, *J. Phys. Soc. Jpn.*, **58**, 301 (1989).
22. K. Edamatsu, T. Nanba and M. Ikezawa, *J. Phys. Soc. Jpn.*, **58**, 314 (1989).
23. G.L. Bottger and A.L. Geddes, *Spectrochimi. Acta Part A*, **23**, 1551 (1967).
24. R.W. Keyes, *Phys. Rev.*, **92**, 580 (1953).
25. C. Kaneta and A. Morita, *J. Phys. Soc. Jpn.*, *55*, 1224 (1986).
26. M. Ikezawa, Y. Kondo and I. Shirotani, *J. Phys. Soc. Jpn.*, **52**, 1518 (1983).
27. S. Sugai and I. Shirotani, *Solid State Commun.*, **53**, 753 (1985).
28. S. Terada, T. Hattori, M. Kobayashi, Y. Akahama, S. Endo and S. Narita, *J. Phys. Soc. Jpn.*, **52**, 2630 (1983).
29. H. Asahina and A. Morita, *J. Phys. C*, **17**, 1839 (1984).
30. G. Lucovsky and R.M. Martin, *Phys. Rev. B*, **4**, 1367 (1971).
31. A. Brown and S. Rundqvist, *Acta Crystallogr.*, **19**, 684 (1965).
32. I.V. Berman and N.B. Brand, *JETP Lett.*, **7**, 323 (1968).
33. J. Wittig and B.T. Matthias, *Science*, **160**, 994 (1968).
34. T. Kikegawa and H. Iwasaki, *Acta Crystallogr. Sect B*, **39**, 153 (1983).
35. M. Ikezawa and T. Nanba, *J. Phys. Soc. Jpn.*, **45**, 148 (1978).
36. G. Nelin and G. Nilsson, *Phys. Rev. B*, **10**, 612 (1974).

4

Far-Infrared Double Polarization Modulation Spectroscopy and Its Application to Reflection–Absorption Spectroscopy

LAWRENCE B. COLEMAN
Department of Physics, University of California at Davis, Davis, CA 95616, USA

[Dedicated to the memory of Alice Valene Coleman]

CONTENTS

Principles and Applications of Polarization-Division Interferometry
Edited by P. L. Polavarapu

1 INTRODUCTION

Double modulation techniques have become standard among the tools used in spectroscopy when the signal of interest is small in comparison with the source signal. To be effective not only must the modulations used produce a detectable change in the detected signal but also there must be a direct and easily determined link to a physical property of interest. Polarization modulation has been used for many years in systems where either inherent orientational order or externally induced changes in the structure produce polarization-specific spectra. Polarization modulation techniques have been most successful in vibrational circular dichroism (VCD), vibrational linear dichroism (VLD) and infrared reflection–absorption spectroscopy (IRRAS). Many laboratories have successfully employed these types of spectroscopy in the visible and infrared spectral regions; however, extending polarization modulation into the very far-infrared region is difficult. This chapter describes the theory, construction and operation of a far-infrared Fourier transform double polarization modulation spectrometer and its application to far-infrared reflection– absorption spectroscopy.

2 POLARIZATION AND DOUBLE POLARIZATION MODULATION IN THE INFRARED

Modern Fourier transform spectrometers based on the Michelson interferometer have revolutionized the study of infrared spectroscopy with their ease of operation, high photometric accuracy and sensitivity and advances in Fourier transform infrared (FTIR) instrumentation have permitted the study of increasingly difficult systems. Fourier transform spectrometers can be categorized by the manner in which data are gathered as the mirror is moved and how the wave fronts are divided. The majority of commercially available FTIR spectrometers use continuous motion of the moving mirror and an amplitude division of the light using a dielectric beamsplitter. The signal from such spectrometers is amplitude modulated by the rapid change in path difference as one mirror is moved. Dielectric beamsplitters are chosen for the frequency range under study, typically potassium bromide for the infrared and Mylar films for the far infrared. Polarization modulation in the infrared is accomplished by the addition of a photoelastic modulator (PEM) and polarizer [1–3]. The technique is very effective in the infrared where the amplitude modulation of the scanning mirror and the natural frequency of the PEM are both in the 10–100 KHz frequency range. This amplitude–polarization modulation combination has been successful in dichroic measurements and in reflection-absorption spectroscopy of films as thin as a monolayer.

Martin and Puplett [4, 5] modified the classical Michelson interferometer to employ polarization division of the light. This spectrometer is described briefly below and in detail in Chapter 1. Recently, Polavarapu and colleagues have utilized the Martin–Puplett interferometer (MPI) design and continuous scanning to create a polarization modulation spectrometer [6–10]. This design, described in detail in Chapter 5, has been successful in the mid- and far-infrared spectral regions.

There are several factors that have limited the extension of polarization modulation into the far-infrared. Most attempts have involved use of rapid scan interferometers and PEMs. Because the polarization modulation frequency must be greater than the amplitude modulation frequency (determined by the mirror scan speed), the typical frequencies of the PEMs (100 kHz range) are ideal. However, the usable frequency response range of the cryogenic detectors used in the far-IR region is typically between 30 and 300 Hz. While the mirror scanning speed can be reduced sufficiently to allow the bolometer to respond, the rapid scan drives can be unstable at these low speeds. In addition, the most commonly used PEM material, zinc selenide, is opaque below 400 cm^{-1}. Griffiths *et al.* [11] attempted to work around the inherent high frequency of the PEM systems by beating two high-frequency (~140 kHz) signals together to create a low frequency (1 kHz) signal containing the polarization-modulated information (a triple modulation spectrometer). The lowest frequency with which data could be obtained with this technique was 700 cm^{-1}.

Another problem lies with the amplitude division process of the interferometer itself. The reflectance of the dielectric beamsplitters is dependent on the polarization in addition to the angle of incidence. The commonly used Mylar beamsplitter, with an angle of incidence of 45°, has a reflectance that is 3.6 times greater for s- than for p-polarized radiation. This polarization dependence of the beamsplitter efficiency creates a large non-interferometric (DC) component in the polarization modulated signal.

3 FAR-INFRARED DOUBLE POLARIZATION MODULATION SPECTROMETER

Our spectrometer design is based on the Martin–Puplett interferometer (MPI), which has been described in detail by Martin [4, 5]. In order to discuss our double modulation design, we first briefly describe the theory and operation of the MPI.

3.1 THE MARTIN–PUPLETT INTERFEROMETER

The basic configuration of the MPI spectrometer is shown in Figure 1. The key elements of the MPI are an input polarizer (P_{in}), an output polarizer (P_{out}), a

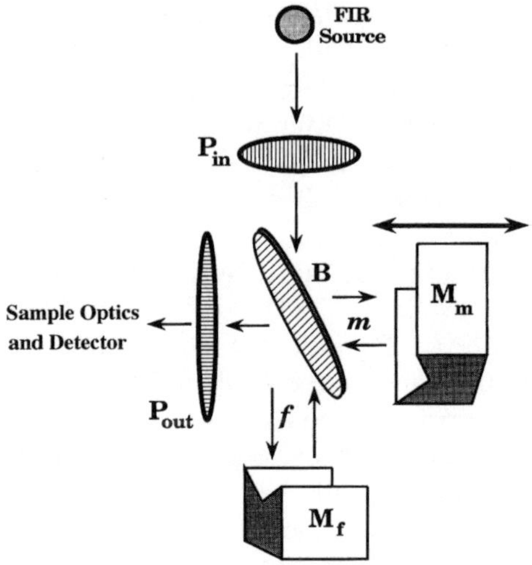

Figure 1 Schematic diagram of the Martin–Puplett interferometer (MPI). P_{in} and P_{out} are polarizers, B is a wire grid polarizer and M_m and M_f are 90° roof-top mirrors. P_{in} modulates the polarization and M_m is moved creating the interferogram

polarizing beamsplitter (B) and two roof-top mirrors (M_m and M_f). The beamsplitter is typically a free-standing fine wire grid. These grids have been shown to be effective polarizers for wavelengths large in comparison with the wire spacing [5, 12, 13]. Wire grid polarizers are commercially available [14]. The grids reflect the component of the incident radiation parallel to the direction of the wires and transmit the component of the incident radiation normal to the direction of the wires. The roof-top mirrors are constructed of two flat mirrors joined at 90°. Light from an unpolarized source is passed through the input polarizer, oriented to produce a beam with the plane of polarization normal to the page. The beamsplitter is oriented at an angle of 45° with respect to the input polarizer. In this configuration, the beamsplitter equally divides the incident polarized radiation—one polarization component into arm *m* and the other component into arm *f*. Upon reflection, the roof-top mirrors rotate the plane of polarization by 90°. In arm *m*, after reflection from the moveable roof-top mirror M_m, the radiation that initially was reflected by the beamsplitter is transmitted. In arm *f* after reflection from the fixed mirror M_f, the light is completely reflected by the wire grid. The two beams recombine on leaving the beamsplitter. As the moving mirror is scanned, the phase difference between the two linearly polarized beams creates an elliptically polarized output beam. The direction of the major axis of the ellipse is determined by the path difference between the two arms, and rotates as M_m is translated.

When the moveable mirror M_m is positioned such that paths *a* and *b* are of equal length (zero path difference position), the recombined radiation will have the same polarization as the radiation originally incident on the beamsplitter. Thus, when the polarization axis of P_{out} is parallel to P_{in}, the beam is totally transmitted. For a monochromatic source, as the path difference changes, the major axis of the recombined beam rotates and the amount of power transmitted by the output polarizer will vary periodically with the path difference. For a continuous spectral source, the power in each spectral interval will be coded differently as the path difference is varied and the output power will yield an interferogram typical of a Michelson interferometer (Figure 2).

Depending on the relative orientations of the input and output polarizers, two different types of interferograms are created. When the orientations of the input and output polarizers are parallel and the moveable mirror M_m is at zero path difference, each spectral component of the radiation transmitted though P_{in} will also be transmitted through P_{out}, and therefore the output intensity will be at a maximum. As M_m is moved, each spectral component of the beam will have its polarization axis oriented differently with respect to the axis of P_{out} and therefore the output intensity will vary with path difference. This results in a 'normal' interferogram. For orthogonal polarizers the intensity of the output beam will be zero at zero path difference and as mirror M_m is translated, the signal will approach one half the total intensity. This creates a complementary interferogram (Figure 2).

As noted by Martin [5], the four different configurations of input and output polarizers and the four interferograms that they produce are a result of the four-port character of the interferometer. The two orthogonal polarization states of the input polarizer can be thought of as two orthogonally polarized input ports, and correspondingly the two orthogonal polarization states of the output polarizer as two orthogonally polarized output ports. By modulating the input

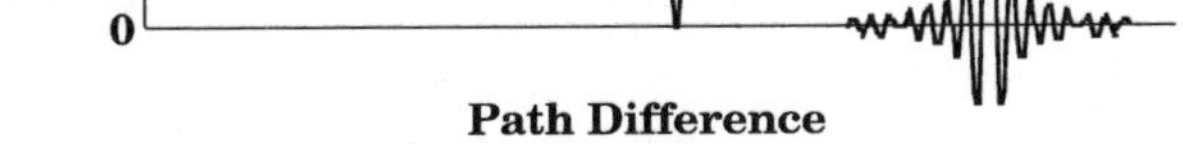

Figure 2 Interferograms produced by the MPI. The i ('normal') interferogram results from parallel polarizers, the c ('complementary') interferogram results from orthogonal polarizers. i − c is the 'difference' interferogram resulting from switching the input polarizer. The difference interferogram is the typical MPI output

port (beam) with a fixed setting of the output polarizer, the two orthogonal orientations of the input polarizer produce a complementary pair of interferograms. The resulting signal is the difference between the 'normal' and the 'complementary' interferograms. Examining Figure 2, we see that this difference interferogram eliminates the mean level and increases the features that carry the spectral information. Thus the MPI produces interferograms with an enhanced dynamic range and sensitivity.

The MPI can be operated with either the input or output polarizer serving the modulation function.

Martin [5] and Carli and Mencaraglia [15] have presented formal Jones matrix analyses [16] of the MPI. The result of that more detailed analysis demonstrates that in the single modulation mode, the perfect MPI delivers half of the input intensity to the detector with a mean signal level of zero.

3.2 THE DOUBLE POLARIZATION MODULATION MPI

For double polarization modulation, both the input and output polarizers are switched and thus the double polarization modulation MPI (DPM–MPI) spectrometer takes advantage of all four ports of an MPI simultaneously. We first present a simple description of the operation of the double modulation spectrometer by considering the interferograms created in each polarizer configuration. A more rigorous analysis of the DPM–MPI follows below.

The input and output polarizers are each switching between two orthogonal polarization directions labeled s and p. These are the two orthogonal directions—in the plane of (p) and perpendicular to (s) the roof-top mirror edges. The switching frequencies, f_1 and f_2 are chosen so that f_2 is at least twice f_1. Four different interferograms are created as the input and output polarizers are switched. These interferograms, tabulated in Figure 3, are labeled i_s, i_p, c_s and c_p, where i denotes a normal interferogram and c denotes a complementary interferogram and the subscripts s and p refer to the orientation of the transmission axis of the output polarizer. As in the MPI when the input and output polarizers axes are parallel, the output interferogram is normal. When they are perpendicular, the output interferogram is complementary.

With the input polarizer in one of the two polarization states, and the output polarizer switching between the s and p orientations, the signal at f_1 will be the difference between the average interferograms of each orientation of P_{in}. This can be written as:

$$\text{signal } (f_1) = \tfrac{1}{2}(i_p + c_s) - \tfrac{1}{2}(c_p + i_s) \quad (1)$$

where the first term in parentheses represents the signal for P_{in} oriented p and the second for P_{in} oriented s. With some rearranging we obtain

$$\text{signal } (f_1) = \tfrac{1}{2}(i_p - c_p) - \tfrac{1}{2}(i_s - c_s) \quad (2)$$

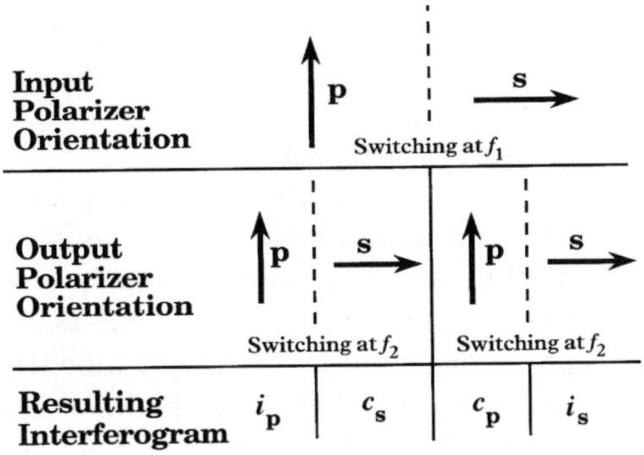

Figure 3 The creation of the four interferograms produced by the configurations of the two modulating polarizers of the DPM–MPI

In equation 2, each term in parentheses is an MPI interferogram (the normal interferogram minus the complementary interferogram). The first term contains interferograms with p-polarization and the second only with s-polarization, and consequently the actual recorded signal from P_{in} will be the difference between these two polarized interferograms. For the ideal spectrometer with no polarization dependent absorbing sample in the beam, this difference is zero. However, when a sample is placed in the output beam, any polarization dependence of the interaction of the beam with the sample will be measured. In other words, the signal at f_1 is the difference between the intensities of the output beam in the p and s orientations $(I_p - I_s)$.

We now consider the f_2 signal. The signal at f_2 will be the difference between the s and p positions of the output polarizer. Since the input polarizer is also switching between the s and p polarization states, the recorded signal will average these two cases, and the recorded signal will be:

$$\text{signal } (f_2) = \tfrac{1}{2}(i_p - c_s) + \tfrac{1}{2}(c_p - i_s) \tag{3}$$

and with some rearranging this becomes

$$\text{signal } (f_2) = \tfrac{1}{2}(i_p + c_p) - \tfrac{1}{2}(i_s + c_s) \tag{4}$$

Both terms in equation 4 are the sum of the normal and complementary interferograms. These sums are the mean level of the interferogram for all path differences and thus they contain no spectral information and consequently the signal synchronous with the output polarizer gives a null result.

The desired sum signal would be obtained if the two terms in equation 3 were subtracted instead of added. This sign change can be created electronically by inverting the signal at f_2 every half cycle of f_1 at the time the input polarizer P_{in} switches polarization. The f_2 signal is then

$$\text{signal } (f_2) = \tfrac{1}{2}(i_p - c_s) - \tfrac{1}{2}(c_p - i_s) \tag{5}$$

or

$$\text{signal } (f_2) = \tfrac{1}{2}\,(i_p - c_p) + \tfrac{1}{2}\,(i_s - c_s) \tag{6}$$

This is the sum of the polarized interferograms ($I_p + I_s$).

Hence if a polarization dependent sample is placed in the output beam of the MPI in this double modulation mode, both the difference and the sum of the orthogonal sample absorptions will be recorded simultaneously. The difference interferogram arises from the input polarizer modulation and the sum interferogram arises from the output polarizer modulation. This double modulation scheme differs from those described by Polavarapu and co-workers [6, 8, 10], Nafie and Diem [1, 2], Golden and co-workers [3, 17–19] and others for use in the infrared region in that the modulation of both channels is a polarization modulation.

4 DETAILED ANALYSIS OF THE DPM–MPI

4.1 JONES MATRIX ANALYSIS

A Jones matrix analysis of the double polarization modulation interferometer (following the analysis of the MPI by Martin) has been presented by DaCosta and Coleman [20, 21]. Recent work in our laboratory by Skolones and Coleman [22, 23] to extract more information from the double polarization modulation interferometer by using digital signal processing techniques has led to a more detailed Jones matrix analysis, including particular attention to the frequency dependence of the detector signal. We present here the core of that analysis.

We define J_{in} and J_{out} be the Jones vectors of a monochromatic beam leaving the input and output polarizers and M_{op} and M_{int} as the Jones matrices corresponding to the output polarizer and MPI interferometer, respectively. The relationship between J_{out} and J_{in} is then

$$J_{out} = M_{op} M_{int} J_{in} \tag{7}$$

If θ_{in} is the angle of the transmission axis of the input polarizer with respect to the vertical, the input beam takes the form

$$J_{in} = \begin{bmatrix} S \\ P \end{bmatrix}_{in} = \begin{bmatrix} A_s \cos(\theta_{in}) \\ A_p \sin(\theta_{in}) \end{bmatrix} \tag{8}$$

For the ideal spectrometer and unpolarized source, the input amplitudes A_s and A_p are equal.

The interferometer matrix represents the interaction of the light beam with the parallel-wire polarization beamsplitter and the roof-top mirrors, which upon reflection rotate the polarization 90°. As shown by DaCosta and co-workers [20–23], the interferometer matrix can be written as

$$M_{\text{int}} = \frac{1}{2}\begin{bmatrix} d_+ & d_- \\ -d_- & -d_+ \end{bmatrix} \tag{9}$$

with d_+ and d_- defined as

$$d_+ = d_\text{f} + d_\text{m} \qquad \text{and} \qquad d_- = d_\text{f} - d_\text{m} \tag{10}$$

and d_f and d_m are the complex propagation coefficients for light traveling through the interferometer via the fixed and moveable mirror arms. The interferometric modulation arises from the difference in phase (Δ) between these two paths. For the perfectly aligned interferometer we obtain

$$\left.\begin{aligned} d_+^2 &= |d_\text{f} + d_\text{m}|^2 = |d_\text{f}|^2 + |d_\text{m}|^2 + 2|d_\text{f}||d_\text{m}|\cos\Delta \\ d_-^2 &= |d_\text{f} - d_\text{m}|^2 = |d_\text{f}|^2 + |d_\text{m}|^2 - 2|d_\text{f}||d_\text{m}|\cos\Delta \\ d_+d_-^* &= |d_\text{f}|^2 - |d_\text{m}|^2 - 2i|d_\text{f}||d_\text{m}|\sin\Delta \\ d_-d_+^* &= |d_\text{f}|^2 - |d_\text{m}|^2 + 2i|d_\text{f}||d_\text{m}|\sin\Delta \end{aligned}\right\} \tag{11}$$

Also from DaCosta and co-workers, the Jones matrix representing the output polarizer is

$$M_{\text{op}} = \begin{bmatrix} \cos^2(\theta_{\text{out}}) & -\sin(\theta_{\text{out}})\cos(\theta_{\text{out}}) \\ -\sin(\theta_{\text{out}})\cos(\theta_{\text{out}}) & \sin^2(\theta_{\text{out}}) \end{bmatrix}$$

with the angle of the output polarizer transmission axis also measured with respect to the vertical. Equation 7 becomes

$$J_{\text{out}} = \frac{1}{2}\begin{bmatrix} \cos^2(\theta_{\text{out}}) & -\sin(\theta_{\text{out}})\cos(\theta_{\text{out}}) \\ -\sin(\theta_{\text{out}})\cos(\theta_{\text{out}}) & \sin^2(\theta_{\text{out}}) \end{bmatrix}\begin{bmatrix} d_+ & d_- \\ -d_- & -d_+ \end{bmatrix}\begin{bmatrix} A_\text{s}\cos(\theta_{\text{in}}) \\ A_\text{p}\sin(\theta_{\text{in}}) \end{bmatrix}$$

Performing the multiplication and simplifying gives

$$J_{\text{out}} = \frac{1}{2}\begin{bmatrix} A_s^2(\cos^2(\theta_{\text{out}})(d_+\cos(\theta_{\text{in}}) + d_-\sin(\theta_{\text{in}})) \\ A_p^2(-\sin(\theta_{\text{out}})\cos(\theta_{\text{out}})(d_+\cos(\theta_{\text{in}}) + d_-\sin(\theta_{\text{in}})) \end{bmatrix}$$
$$\begin{matrix} + \sin(\theta_{\text{out}})\cos(\theta_{\text{out}})(d_-\cos(\theta_{\text{in}}) + d_+\sin(\theta_{\text{in}}))) \\ - \sin^2(\theta_{\text{out}})(d_-\cos(\theta_{\text{in}}) + d_+\sin(\theta_{\text{in}}))) \end{matrix}\Bigg] \tag{12}$$

from this expression, we can find the intensity of the S- and P-polarized components of the output light beam;

$$\begin{aligned}S^2 = \tfrac{1}{4} A_s^2 \cos^2(\theta_{out})[(d_+^2 \cos^2(\theta_{in}) + (d_+d_-^* + d_-d_+^*)\sin(\theta_{in})\cos(\theta_{in}) \\ + d_-^2 \sin^2(\theta_{in}))\cos^2(\theta_{out}) + ((d_+d_-^* + d_-d_+^*)(\cos^2(\theta_{in}) + \sin^2(\theta_{in})) \\ + 2(d_+^2 + d_-^2)\sin(\theta_{in})\cos(\theta_{in}))\sin(\theta_{out})\cos(\theta_{out}) + (d_-^2 \cos^2(\theta_{in}) \\ + (d_+d_-^* + d_-d_+^*)\sin(\theta_{in})\cos(\theta_{in}) + d_+^2 \sin^2(\theta_{in}))\sin^2(\theta_{out})]\end{aligned}$$

and

$$\begin{aligned}P^2 = \tfrac{1}{4} A_p^2 \sin^2(\theta_{out})[(d_+^2 \cos^2(\theta_{in}) + (d_+d_-^* + d_-d_+^*)\sin(\theta_{in})\cos(\theta_{in}) \\ + d_-^2 \sin^2(\theta_{in}))\cos^2(\theta_{out}) + ((d_+d_-^* + d_-d_+^*)(\cos^2(\theta_{in}) + \sin^2(\theta_{in})) \\ + 2(d_+^2 + d_-^2)\sin(\theta_{in})\cos(\theta_{in}))\sin(\theta_{out})\cos(\theta_{out}) + (d_-^2 \cos^2(\theta_{in}) \\ + (d_+d_-^* + d_-d_+^*)\sin(\theta_{in})\cos(\theta_{in}) + d_+^2 \sin^2(\theta_{in}))\sin^2(\theta_{out})] \qquad (13)\end{aligned}$$

The terms in brackets are identical and contain all the interferometric information. We refer to this term as the interferometer term, B.

The total intensity of the beam leaving the output polarizer is

$$I_{out} = S^2 + P^2 = \tfrac{1}{4} B[A_s^2 \cos^2(\theta_{out}) + A_p^2 \sin^2(\theta_{out})] \qquad (14)$$

With some rearranging, we obtain

$$\begin{aligned}B = d_+^2[\cos^2(\theta_{in})\cos^2(\theta_{out}) + \sin^2(\theta_{in})\sin^2(\theta_{out})] \\ + d_-^2[\sin^2(\theta_{in})\cos^2(\theta_{out}) + \cos^2(\theta_{in})\sin^2(\theta_{out})] \\ + 2(d_+^2 + d_-^2)\sin(\theta_{in})\cos(\theta_{in})\sin(\theta_{out})\cos(\theta_{out}) \\ + (d_+d_-^* + d_-d_+^*)\sin(\theta_{in})\cos(\theta_{in})[1 + \sin(\theta_{out})\cos(\theta_{out})] \qquad (15)\end{aligned}$$

Substituting the expressions for $d_\pm^2$ and $d_-d_+^*$ (equations 11) and rearranging gives

$$\begin{aligned}B = (d_f^2 + d_m^2)[(\cos^2(\theta_{in})\cos^2(\theta_{out}) + \sin^2(\theta_{in})\sin^2(\theta_{out})) + (\sin^2(\theta_{in})\cos^2(\theta_{out}) \\ + \cos^2(\theta_{in})\sin^2(\theta_{out})) + 4 \sin(\theta_{in})\cos(\theta_{in})\sin(\theta_{out})\cos(\theta_{out})] \\ + 2 \,|d_f||d_m|\, \cos\Delta[(\cos^2(\theta_{in})\cos^2(\theta_{out}) + \sin^2(\theta_{in})\sin^2(\theta_{out})) \\ - (\sin^2(\theta_{in})\cos^2(\theta_{out}) + \cos^2(\theta_{in})\sin^2(\theta_{out}))] \\ + 2(d_f^2 - d_m^2)\sin(\theta_{in})\cos(\theta_{in})(1 + \sin(\theta_{out})\cos(\theta_{out})) \qquad (16)\end{aligned}$$

For an ideal instrument $|d_f| = |d_m| = 1$ and the interferometer term reduces to

$$\begin{aligned}B = 2[(\cos^2(\theta_{in})\cos^2(\theta_{out}) + \sin^2(\theta_{in})\sin^2(\theta_{out})) + (\sin^2(\theta_{in})\cos^2(\theta_{out}) \\ + \cos^2(\theta_{in})\sin^2(\theta_{out})) + 4\sin(\theta_{in})\cos(\theta_{in})\sin(\theta_{out})\cos(\theta_{out})] \\ + 2 \cos\Delta[(\cos^2(\theta_{in})\cos^2(\theta_{out}) + \sin^2(\theta_{in})\sin^2(\theta_{out})) \\ - (\sin^2(\theta_{in})\cos^2(\theta_{out}) + \cos^2(\theta_{in})\sin^2(\theta_{out}))]\end{aligned}$$

The first term is independent of the phase difference Δ, and so the interferometric modulation is carried solely by the second term. Using the identities

$$\left.\begin{aligned} 2\cos^2\theta &= \cos 2\theta + 1 \\ 2\sin\theta\cos\theta &= \sin 2\theta \\ 2\sin^2\theta &= 1 - \cos 2\theta \end{aligned}\right\} \tag{17}$$

we find

$$B = 2[1 + \sin(2\theta_{\text{in}})\sin(2\theta_{\text{out}}) + \cos\Delta\cos(2\theta_{\text{in}})\cos(2\theta_{\text{out}})] \tag{18}$$

The exit intensity can now be written as

$$\begin{aligned} I_{\text{out}} = S^2 + P^2 &= \tfrac{1}{2}[A_s^2\cos^2(\theta_{\text{out}}) + A_p^2\sin^2(\theta_{\text{out}})] \\ &\times [1 + \sin(2\theta_{\text{in}})\sin(2\theta_{\text{out}}) + \cos\Delta\cos(2\theta_{\text{in}})\cos(2\theta_{\text{out}})] \end{aligned} \tag{19}$$

For the perfect spectrometer with no sample absorption, $A_s^2 = A_p^2 = I_{\text{in}}/2$, where I_{in} is the source intensity entering the input polarizer. The intensity at the detector is then

$$I_{\text{out}} = \frac{I_{\text{in}}}{4}[1 + \sin(2\theta_{\text{in}})\sin(2\theta_{\text{out}}) + \cos\Delta\cos(2\theta_{\text{in}})\cos(2\theta_{\text{out}})] \tag{20}$$

4.2 FREQUENCY DEPENDENT SIGNAL

Examining the three terms of the output intensity, we see that the first term is a constant. The second term varies in time (as the polarizers rotate), but is independent of path length difference—no interferometric modulation. The third term has both a time dependence and interferometric information. We will refer to the second term as I^{t} and the third term as I^{ts}.

To calculate a theoretical frequency-dependent signal that we may visualize graphically, we replace $2\theta_{\text{in}}$ with f and $2\theta_{\text{out}}$ with rf, where r is a positive constant. Then,

$$I_{\text{t}} = \sin(f)\sin(rf)$$

and

$$I_{\text{ts}} = \cos(\Delta)\cos(f)\cos(rf)$$

Figure 4 presents I_{t} and I_{ts}. as functions of frequency for $r = 42/155$ (this ratio represents the two modulation frequencies, 42 and 155 Hz, used in our DPM–MPI spectrometer) and $\cos\Delta = 1$ (zero path length difference, or 'white light'). Figure 5 shows the sum of the two terms and therefore represents the theoretical signal at the detector. As the moving mirror is moved, the phase relationship between the two terms changes and the signal is no longer sinusoidal. Figure 5 also shows the sum $(I_{\text{t}} + I_{\text{ts}})$ for $\Delta = \{\pi/4$ and $\pi/2\}$.

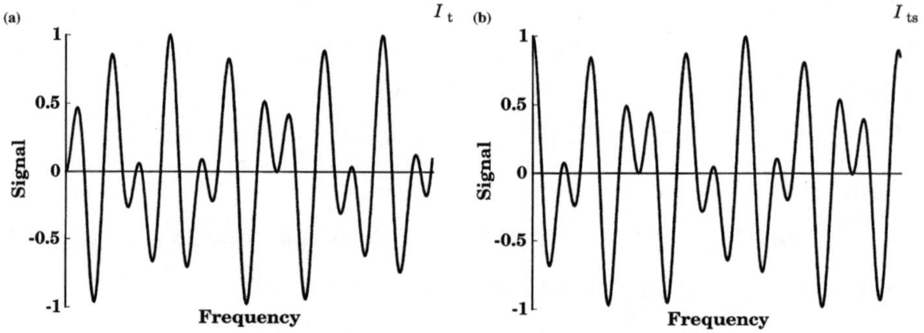

Figure 4 (a) The time-dependent (I_t) and (b) the time- and path difference-dependent (I_{ts}) terms of the DPM–MPI signal at zero path difference ($\Delta = 0$) versus frequency. The modulation frequencies are taken as 42 and 155 Hz. The scales are arbitrary

These results can be more easily understood if the expressions for I_t and I_{ts} are transformed using Euler identities for sine and cosine, giving

$$I_t = -\tfrac{1}{2}[\cos(\theta_{in} + \theta_{out}) - \cos(\theta_{in} - \theta_{out})]$$

and

$$I_{ts} = \tfrac{1}{2}\cos\Delta[\cos(\theta_{in} + \theta_{out}) + \cos(\theta_{in} - \theta_{out})]$$

At zero path length difference, $\cos\Delta = 1$ and we have

$$I_{ts} + I_t = \cos(\theta_{in} - \theta_{out})$$

whereas at $\Delta = \pi$, $\cos\Delta = -1$ and

$$I_{ts} + I_t = -\cos(\theta_{in} + \theta_{out})$$

At white light the signal is a sinusoid whose frequency is the difference of the input and output polarizer frequencies, while for $\Delta = \pi$ the signal is again sinusoidal, but of frequency equal to the sum of the polarizer frequencies. At intermediate path length differences the signal is a combination of these two cases:

$$I_{ts} + I_t = \tfrac{1}{2}[(\cos(\theta_{in} - \theta_{out}) - \cos(\theta_{in} + \theta_{out})) + (\cos(\theta_{in} - \theta_{out}) - \cos(\theta_{in} + \theta_{out}))\cos\Delta] \tag{21}$$

4.3 SIGNAL WITH A DICHROIC SAMPLE

With a dichroic sample in the exit beam, the S- and P-polarized components of the beam will interact differently with the sample. The S- and P-amplitudes are

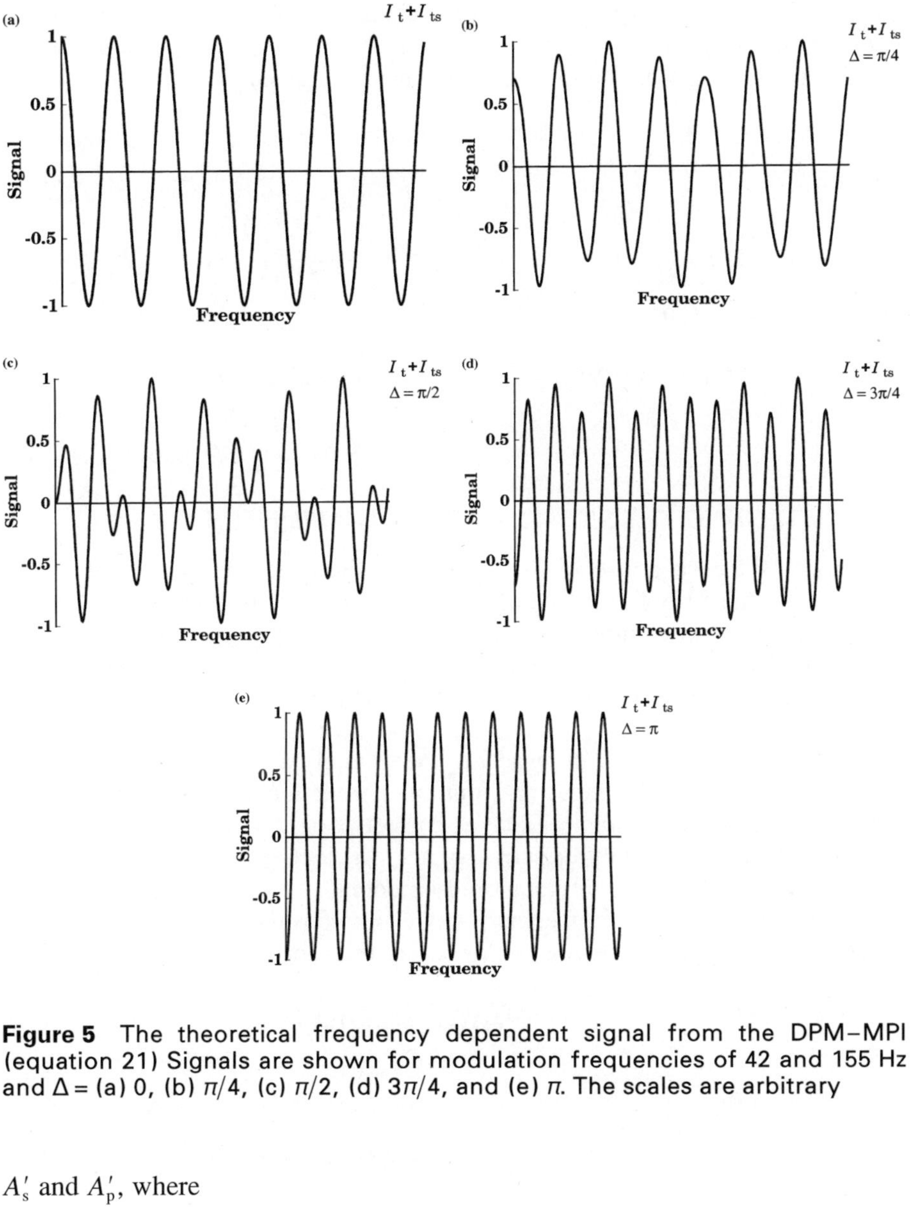

Figure 5 The theoretical frequency dependent signal from the DPM–MPI (equation 21) Signals are shown for modulation frequencies of 42 and 155 Hz and $\Delta =$ (a) 0, (b) $\pi/4$, (c) $\pi/2$, (d) $3\pi/4$, and (e) π. The scales are arbitrary

A'_s and A'_p, where

$$A'_s = T_s A_s$$
$$A'_p = T_p A_p$$

Here T_s and T_p are the sample transmission coefficients for the S- and P-polarization states. In general, T_s and T_p will be frequency dependent. Since we have only considered a monochromatic source, this frequency dependence is not explicitly written.

After interacting with the sample, the intensity is given by

$$I_{\text{sample}} = \frac{I_{\text{in}}}{2}[T_s^2\cos^2(\theta_{\text{out}}) + T_p^2\sin^2(\theta_{\text{out}})][1 + \sin(2\theta_{\text{in}})\sin(2\theta_{\text{out}}) + \cos\Delta\cos(2\theta_{\text{in}})\cos(2\theta_{\text{out}})]$$

Using the trigonometric identities of equations 17, we find

$$T_s^2\cos^2(\theta_{\text{out}}) + T_p^2\sin^2(\theta_{\text{out}}) = \tfrac{1}{2}[(T_s^2 - T_p^2)\cos(2\theta_{\text{out}}) + T_s^2 + T_p^2]$$

and the expression for intensity becomes

$$I_{\text{sample}} = \frac{I_{\text{in}}}{4}[(T_s^2 + T_p^2)\cos(2\theta_{\text{out}}) + T_s^2 + T_p^2][1 + \sin(2\theta_{\text{in}})\sin(2\theta_{\text{out}}) + \cos\Delta\cos(2\theta_{\text{in}})\cos(2\theta_{\text{out}})] \quad (22)$$

4.4 DETECTING THE SIGNAL—LOCK-IN AMPLIFIERS

The goal of the spectroscopy is to measure the coefficients T_s^2 and T_p^2, or equivalently the ratio T_s/T_p. The usual method and that used for the majority of the studies in our laboratory is to use two lock-in amplifiers, one tuned to the input polarizer frequency ω_{in}, where $2\phi_{\text{in}} = \omega_{\text{in}}t \times \phi_{\text{in}}$, and the other tuned to the output polarizer frequency ω_{out} where $2\theta_{\text{out}} = \omega_{\text{out}}t + \phi_{\text{out}}$. For this discussion we assume $\phi_{\text{in}} = \phi_{\text{out}} = 0$. Examining equation 22, we see that the term (cosines) is out of phase with respect to the spatially constant term (sines) by a factor $\pi/2$ for both the input and output modulation frequencies. With the phases of the two lock-in amplifiers set correctly, only the interferometric term will contribute to the amplifier output signal. The lock-in amplifier method thus ignores the pure constant and temporally modulated components of the signal, and extracts only the coefficient of the interferometric term. Expanding equation 22, only the following two terms are relevant:

$$I_1 = \frac{I_{\text{in}}}{4}(T_s^2 - T_p^2)\cos\Delta\cos(\omega_{\text{in}}t)\cos^2(\omega_{\text{out}}t)$$

and

$$I_2 = \frac{I_{\text{in}}}{4}(T_s^2 + T_p^2)\cos\Delta\cos(\omega_{\text{in}}t)\cos(\omega_{\text{out}}t) \quad (23)$$

These represent the difference and the sum of the two transmitted intensities.

Using the trigonometric identities of equation 17, the first term becomes

$$I_1 = \frac{I_{\text{in}}}{8}(T_s^2 - T_p^2)\cos\Delta\cos(\omega_{\text{in}}t) + \frac{I_{\text{in}}}{8}(T_s^2 - T_p^2)\cos\Delta\cos(\omega_{\text{in}}t)\cos(2\omega_{\text{out}}t) \quad (24)$$

This can be transformed using the identity

$$\cos(\theta_{\text{in}})\cos(\theta_{\text{out}}) = \tfrac{1}{2}[\cos(\theta_{\text{in}} + \theta_{\text{out}}) + \cos(\theta_{\text{in}} - \theta_{\text{out}})]$$

into

$$I_1 = \frac{I_{\text{in}}}{8}(T_s^2 - T_p^2)\cos\Delta\cos(\omega_{\text{in}}t + \varphi_{\text{in}}) + \frac{I_{\text{in}}}{16}(T_s^2 - T_p^2)\cos\Delta[\cos((\omega_{\text{in}} + 2\omega_{\text{out}})t$$
$$+ \varphi_{\text{in}} + \omega_{\text{out}}) + \cos((\omega_{\text{in}} - 2\omega_{\text{out}})t + \varphi_{\text{in}} - \varphi_{\text{out}})] \quad (25)$$

where we have now added potential phase difference terms. A lock-in amplifier tuned to extract the component of the signal at frequency ω_{in} will ignore the second term in I_{f} due to the $\pm 2\omega_{\text{out}}$ frequency shift, and so the amplifier output will be proportional to the difference between the squares of the S- and P-transmission coefficients, $T_{\text{s}}^2 - T_{\text{p}}^2$. The output of the ω_{in}-tuned amplifier is thus the difference signal, I_{diff}.

The term I_2 can be similarly transformed to give

$$I_2 = \frac{I_{\text{in}}}{8}(T_s^2 + T_p^2)\cos\Delta[\cos((\omega_{\text{in}} + \omega_{\text{out}})t + \varphi_{\text{in}} + \varphi_{\text{out}})$$
$$+ \cos((\omega_{\text{in}} - \omega_{\text{out}})t + \varphi_{\text{in}} - \varphi_{\text{out}})] \quad (26)$$

As noted above (see equation 4), this component of the signal will be zero at the output polarizer frequency ω_{out}. To extract the quantity $T_{\text{s}}^2 + T_{\text{p}}^2$ using lock-in detection, two techniques are possible, as follows.

Signal inversion

Before passing the signal through the output frequency tuned amplifier, we modify it by taking only the absolute value of the $\cos(\omega_{\text{in}}t + \varphi_{\text{in}})$ term, thus

$$I_2 = \frac{I_{\text{in}}}{4}(T_s^2 + T_p^2)[\cos\Delta\,|\cos(\omega_{\text{in}}t + \varphi_{\text{in}})|\cos(\omega_{\text{out}}t + \varphi_{\text{out}})]$$

The modulation imposed by the input polarizer will then be continuously non-negative, and so if the filter time constant is sufficiently large the amplifier output will be proportional to the sum of the squares of the S- and P-transmission coefficients, $T_{\text{s}}^2 + T_{\text{p}}^2$. This signal modification can be achieved by multiplying the signal by a square wave of frequency ω_{in} and values ± 1, synchronous with the modulation imposed by the input polarizer. This is the method originally implemented by DaCosta and Coleman [21]. A more complete discussion and analysis of this data acquisition method can be found elsewhere.

Frequency-shifted detection

If a lock-in amplifier is tuned to detect the component of the incoming signal at either of the frequencies $\omega_{\text{in}} + \omega_{\text{out}}$ or $\omega_{\text{in}} - \omega_{\text{out}}$, then equation 26 indicates that the amplifier output will be proportional to $T_{\text{s}}^2 + T_{\text{p}}^2$, provided that the detection phase of the amplifier is properly set.

In either case (a) or (b), we refer to the amplifier output as the sum signal, I_{sum}. We can find the ratio $T_{\text{s}}/T_{\text{p}}$ from the sum and difference signals;

$$\frac{T_s^2}{T_p^2} = \frac{I_{\text{sum}} + I_{\text{diff}}}{I_{\text{sum}} - I_{\text{diff}}}$$

4.5 DETECTING THE SIGNAL—FOURIER DEMODULATION

By employing digital signal processing techniques, we can analyze the complete signal by calculating the Fourier components of each of the frequencies in equations 25 and 26.

Reassembling equation 22 using equations 25 and 26, the intensity can be written as

$$\begin{aligned}I_{\text{sample}} = {} & \frac{I_{\text{in}}}{8}\,[2 + \cos((\omega_{\text{in}} - \omega_{\text{out}})t + \varphi_{\text{in}} - \varphi_{\text{out}}) - \cos((\omega_{\text{in}} + \omega_{\text{out}})t + \varphi_{\text{in}} + \varphi_{\text{out}})] \\ & + \frac{I_{\text{in}}}{8}\,(T_s^2 - T_p^2)\cos\Delta\,\cos(\omega_{\text{in}}t + \varphi_{\text{in}}) \\ & + \frac{I_{\text{in}}}{16}\,(T_s^2 - T_p^2)\cos\Delta[\cos((\omega_{\text{in}} + 2\omega_{\text{out}})t + \varphi_{\text{in}} + 2\varphi_{\text{out}}) \\ & \qquad + \cos((\omega_{\text{in}} - 2\omega_{\text{out}})t + \varphi_{\text{in}} - 2\varphi_{\text{out}})] \\ & + \frac{I_{\text{in}}}{8}\,(T_s^2 - T_p^2)\cos\Delta[\cos((\omega_{\text{in}} + \omega_{\text{out}})t + \varphi_{\text{in}} + \varphi_{\text{out}}) \\ & \qquad + \cos((\omega_{\text{in}} - \omega_{\text{out}})t + \varphi_{\text{in}} - \varphi_{\text{out}})]\end{aligned} \quad (27)$$

We can then extract the quantity $T_{\text{s}}^2 + T_{\text{p}}^2$ by calculating the Fourier component corresponding to either $\omega_{\text{in}} + \omega_{\text{out}}$ or $\omega_{\text{in}} - \omega_{\text{out}}$, and the quantity $T_{\text{s}}^2 - T_{\text{p}}^2$ from the Fourier component corresponding to ω_{in}, $\omega_{\text{in}} + 2\omega_{\text{out}}$ or $\omega_{\text{in}} - 2\omega_{\text{out}}$.

Skolones and Coleman [22, 23] have completed a full analysis of this sampling technique and have experimentally verified that with a dichroic sample in the exit optics, the seven terms above appear in the signal spectrum. The calculated Fourier components of the input polarizer modulation frequency and the sum of the input and output polarizer modulation frequencies produce the difference and sum interferograms.

Lock-in amplifiers efficiently enhance the signal at the reference signal, but they cannot compensate for changes in the phase of the detector signal relative to its synchronous reference. This phase tracking is not required in many applications, however, for a real MPI spectrometer with imperfect alignment or components the phase of the detector output signal can depend on the optical path length difference between the fixed and moveable interferometer arms. Since the lock-in amplifier cannot track this phase change, conventional signal demodulation will degrade the data. In addition, Fourier demodulation eliminates the need to set the phases of the two lock-in amplifiers and the signal inverter. Proper choice of these three phases can be difficult to optimize.

Constructing the interferogram directly from the Fourier transform of the detector signal alleviates these problems.

5 THE DOUBLE POLARIZATION SPECTROMETER

We have implemented the double polarization MPI design by modifying a Beckman RIIC FS-720 modular Michelson interferometer with an FS-821 stepping motor/micrometer mirror drive. This spectrometer had earlier been converted to an MPI with a kit manufactured by QMC Instruments. The MPI modifications include the two roof-top mirrors, wire grid beamsplitters (we have used both 12.5 µm diameter tungsten wires on 25 µm centers and 5 µm diameter tungsten wires on 12.5 µm centers) and two rotating polarizer mounts of local design [24] containing photolithographically deposited wire grids on polyethylene substrates (Cambridge Instruments IGP 224). The QMC supplied vibrating reed polarization modulator was replaced with the rotating polarizer design when it was discovered that its operation led to signal noise. This noise arose from fluctuations in the position of the arc within the quartz envelope of the lamp (As described in Ref.24, while the mercury arc source produces unpolarized radiation, the arc plasma wanders within the quartz tube. The vibrating reed modulator mechanically oscillates two orthogonal polarizers in front of the source. These two polarizers each sweep over a separate and limited area of the aperture, the horizontal polarizer covers the left part of the aperture while the vertical polarizer covers the right side of the aperture. As the arc plasma wanders in the source tube variations in radiation intensity across the face of the input polarizer aperture results in changes in the intensity between the two polarizations creating large abrupt changes in the difference interferogram signal.)

In addition to the MPI modifications, the original source box was replaced to allow easier access to the mercury arc source and collimating optics, and the stepping motor–micrometer drive was modified for smoother operation and smaller step size. The rotating input polarizer is located in front of the high-pressure mercury arc source and the rotating output polarizer is placed at the focus of the interferometer output beam, an image of the source at unit magnification. The beam is then passed into the sample chamber and the detector.

The rotating polarizer modulators [24] are constructed to maximize the throughput of the beam and to eliminate any structures in the beam which would add another modulation frequency to the system (for example, any type of spoke or strut whose interruption of the beam would add an amplitude modulation component to the signal). The design is shown in Figure 6. The polarizer is mounted on a large diameter (2.54 cm i.d.) bearing pressed into an aluminum plate. The polarizer–bearing assembly is driven by a permanent magnet DC motor and pulley system using a n-Buna O-ring 'belt'. Incorporated on the edge of the polarizer mount are two chopper blades positioned 180° apart to insure that the frequency of the reference signal is equal to the switching frequency of the polarizer. The chopper blades trigger a photodiode optical switch providing a reference signal.

Choice and control of the two modulation frequencies are critical to the operation of the spectrometer. Beat frequencies must be avoided as they create large fluctuations in output signals. The two frequencies must also be within the efficient response range of the cryogenic detector. An input polarizer frequency of 42 Hz and an output polarizer frequency of 155 Hz were found to be optimal for reduction of beats and maintenance of detector sensitivity. When using Fourier demodulation, the choice of frequencies is not as critical. The two frequencies must only be chosen to separate the five Fourier components (equation 27). Frequency-driven voltage stabilizers were used to maintain the modulation frequencies to within a few tenths of a hertz.

To obtain the sum interferogram, it is necessary to invert the P_{out} signal every time the input modulator, P_{in}, switches polarization. Using the reference signal

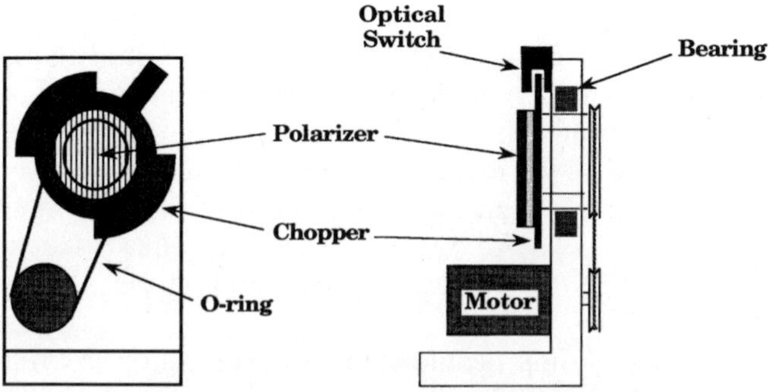

Figure 6 Schematic diagram of the polarization modulator

of P_{in} as a trigger, the output polarizer signal (at f_2) is inverted as the s-orientation input polarizer moves in front of the aperture. The inverter circuitry includes a phase shifter to adjust the inversion timing with respect to the input polarizer position.

A block diagram of the double modulation spectrometer and associated electronics is shown in Figure 7. A regulated DC power supply is necessary for the arc source to eliminate any frequency components in the signal that will interfere or beat with the two modulated signals. The reference frequency of the input modulator, f_i, is used as a reference for both lock-in amplifier A and for the controlled inverter. The far-IR radiation is detected with an Infrared Laboratories liquid helium-cooled composite gallium-doped germanium bolometer. Several bolometers, optimized for different spectral regions, have been used with the spectrometer. The bolometer signal is input into a high-impedance, low noise solid-state preamplifier and the amplified output is split into two channels. One channel is input directly into lock-in amplifier A using reference frequency f_i, and the other channel is input into the controlled inverter, the output of the inverter is input into lock-in amplifier B operating at frequency f_2. The output of amplifier A, the difference interferogram, and the output of amplifier B, the sum interferogram are then digitized through separate channels of an A/D for computer input and processing. To date the majority of our use of the DPM–MPI has been with the analog lock-in signal acquisition system described here and below.

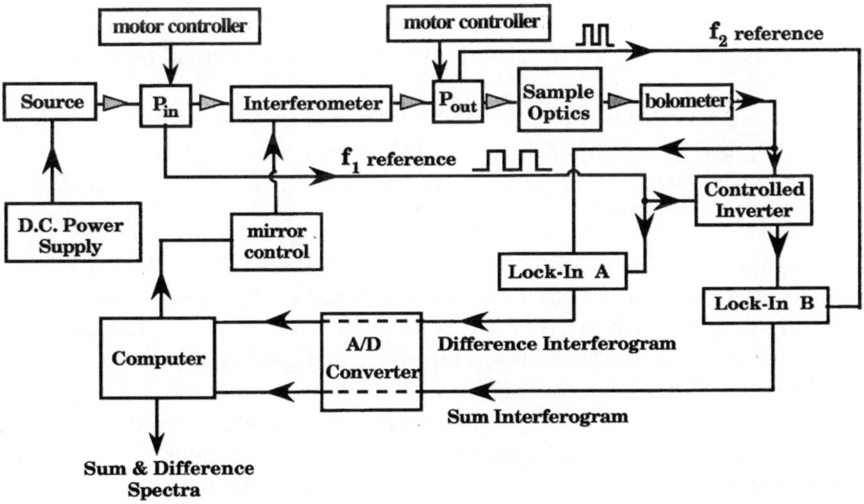

Figure 7 Schematic diagram of the DPM–MPI spectrometer with lock-in amplifier signal demodulation. For Fourier demodulation the signal from the detector is amplified, sampled by the A/D electronics and input directly to the computer

6 SPECTROMETER OPERATION

The interferometer is controlled and the data are acquired and manipulated by a desktop microcomputer using locally written software. At each position of the moving mirror both of the channels of the A/D (outputs of the amplifiers at f_1 and f_2) are sampled. Both signal channels are sampled in a single sweep, which insures that all the points of both interferograms are recorded 'simultaneously'. Multiple sweeps are stored and averaged for each mirror position. The computer software performs real-time Fourier transforms of either the sum of difference interferograms every 2^n points. This allows the user to display either the sum or difference interferogram, or the sum or difference spectrum, during the course of the scan. After the scan, the interferograms are separately stored on disk and the Fourier transformed spectra are plotted.

Analog data acquisition with the DPM–MPI proceeds as follows:

1. The moveable mirror is moved to near the zero path difference position.
2. With the input polarizer rotating at f_1, the output polarizer is positioned with its polarization axis along the p-direction. The spectrometer is then in the normal single modulation MPI mode with an output beam polarized along p.
3. The phase of amplifier A is adjusted so that the out-of-phase signal is zero and the in-phase output signal is *positive*. This insures when in the double modulation mode that the recorded difference interferogram is proportional to $I_p - I_s$.
4. When the output polarizer rotates, amplifier A goes from measuring a large signal proportional to (I_p), to a small difference signal proportional to $I_p - I_s$.
5. The signal inverter is then turned on to provide the sum signal for amplifier B. The phase of amplifier B is adjusted to give a positive signal and a zero out-of-phase signal. The sum signal is then maximized by adjusting the phase of the inverter.

Other aspects of the experiment, such as cut-off filtering, symmetrization of the interferogram and handling of the data pre- and post-Fourier transformation, are similar to practices in single modulation operation of the spectrometer.

With M_m at the position of zero path, the sum signal is a positive maximum value. For the ideal spectrometer, the DC component of the difference interferogram is zero. For sample absorption primarily along the p direction, the difference signal will be negative, conversely, if the sample absorption is primarily along the s-direction, the difference signal will be positive. With no sample in the beam, the difference interferogram would, in theory, be zero. In our current instrument, the difference interferogram signal at zero path position is approximately 3% of the sum interferogram. This residual interferogram arises from linear dichroism in the output optics of the interferometer, dichroism in the polymeric windows and filters in the detector cryostat and a residue of polarized radiation in the interferometer.

7 APPLICATION OF THE DPM–MPI SPECTROMETER

7.1 FAR-INFRARED REFLECTION-ABSORPTION SPECTROSCOPY (FIRRAS)

The DPM–MPI spectrometer described above produces high-quality spectra from 10 to near 350 cm^{-1} where the wire grid beamsplitter loses efficiency. By measuring the spectral differences between two orthogonal polarization states, the DPM–MPI can be used to study a variety of systems.

Our primary use of the DPM–MPI spectrometer has been to study thin films using reflection-absorption spectroscopy. For a dielectric film deposited on a metal substrate, Berreman [25] found that at non-normal incidence, not only does one obtain an absorption at the LO mode frequency, but also the TO mode absorption is practically eliminated from the spectrum when the film thickness is much less the wavelength of the incident radiation. This effect was later noted by Greenler [26] and is the basis for the surface specificity of reflection–absorption spectroscopy. The high conductivity of the metal substrate prohibits a parallel component of the electric field from existing near the surface but enhances the component of the electric field perpendicular to the surface with the result that the incident radiation can couple to vibrational modes with dipoles perpendicular, but not parallel, to the film surface. This has become known as the surface dipole selection rule. In 1966, Greenler [26] considered the interaction of IR radiation with adsorbed molecules on a metal surface and demonstrated that at grazing angles of incidence (>85°) the fields at the surface perpendicular to the plane of incidence (s-polarized) were near zero, owing to the 180° phase shift upon reflection, whereas the fields in the plane of incidence (p-polarized) were enhanced by the 90° phase shift. With one polarization interacting with the surface film and the other showing only the substrate, the double polarization modulation difference spectra produce the typical sample to background ratio spectra, this with significant dynamic range as the background reflectivity and instrumental lineshape of the spectrometer are removed by the polarization modulation. In the infrared region this technique is referred to by some researchers as IR reflection–absorption spectroscopy (IRRAS or RAIRS) or grazing incidence reflectance GIR).

This polarization specificity also leads to the ability to study molecular orientation on the surface, as only the surface normal component of a bond polarizability will couple to the field. In examining crystalline films, the 'surface selection rule' leads to the excitation of the longitudinal optical phonon. Additional quantitative detail of the RA electrodynamics and discussion of spectral modeling have been presented by many researchers [2, 11, 27–30] and by our laboratory [20, 23, 31–34].

By combining IRRAS with high frequency polarization modulation using photoelastic modulators (PEMs), many film and surface coating systems have been studied, including sub-monolayer films [27, 35], working electrochemical electrodes [36], and Langmuir–Blodgett monolayer films [19]. This double modulation IRRAS (DM-IRRAS) has been very successful in the infrared region (4000–500 cm^{-1}); however, as discussed above, successful extension of DM-IRRAS into the far-infrared region has not been routinely available.

7.2 IMPLEMENTATION OF FIRRAS USING THE DPM–MPI

The FIRRAS measurements are performed with the sample optics shown in Figure 8. The sample chamber optical components consist of two aluminized front surface flat mirrors M_1 and M_3 and a spherical mirror M_2 with a focal length of 33 cm. The incident angle of the radiation on the sample mirror is approximately 85°. A brass light pipe (i.d. 1.27 cm), with a light cone attached to one end and a polyethylene vacuum window on the other, is used to guide the reflected beam from the sample mirror on to the detector window. The light pipe partially scrambles the polarization of the exiting beam, which makes the difference spectrum less sensitive to the dichroism of the windows and the detector filters.

The sample chamber is evacuated to below 5×10^{-5} Torr using a mechanical pump-backed diffusion pump and a liquid nitrogen cooled trap. The sample

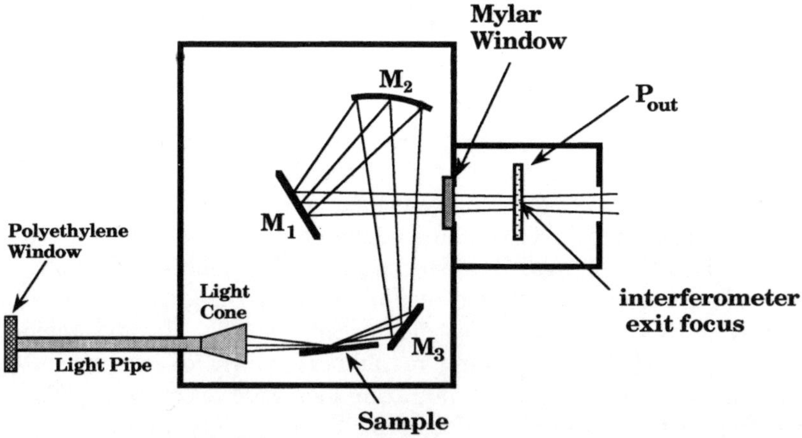

Figure 8 Schematic diagram of the reflection–absorption sample apparatus. The output of the interferometer is focused at the output polarizer. Mirrors M_1 and M_3 steer the beam refocused by spherical mirror M_2 on to the sample at grazing angle

optics are aligned using a He–Ne laser to trace the optical path from the exit window to the arc source. The size of the laser spot on the surface of the sample is used to determine the angle of incidence, ϕ_0. For the spectra presented here ϕ_0 was $85 \pm 2°$. At greater angles, significant amounts of source radiation miss the sample mirror, greatly reducing the signal-to-noise ratio.

A single sample holder was built to control the temperature of the sample from 90 to 420 K.

From the DPM scheme, sum and difference signals are derived from

$$I_\Sigma = I_s + I_p$$
$$I_\Delta = I_s - I_p$$

The reflection-absorption (RA) spectrum is the ratio of the signal with absorption (p-polarization intensity) divided by reflected signal (s-polarization intensity). This is obtained from the sum and difference spectra as

$$\frac{I_\Sigma - I_\Delta}{I_\Sigma + I_\Delta} = \frac{(I_s + I_p) - (I_s - I_p)}{(I_s + I_p) + (I_s - I_p)} = \frac{I_p}{I_s} = \text{the RA spectrum}$$

Further, this ratio can be expressed in terms of the absorptions by each polarization.

$$\text{RA} = \frac{I_p}{I_s} = \frac{1 - A_p}{1 - A_s}$$

where A is the fraction of incident light not reflected. The reflected intensities would be undiminished for a bare conducting substrate and are reduced by film/substrate absorption in either polarization. From these data and detailed modeling of the electrodynamics of the RA process [31–34], the frequency-dependent complex dielectric constant of the film can be determined.

7.3 FIRRAS SPECTRA OF POTASSIUM IODIDE

To demonstrate the capabilities of the DPM–MPI spectrometer, we present FIRRAS spectra of ultra-thin films of potassium iodide.

Polycrystalline films of KI were produced by thermal evaporation on to aluminized glass slide substrates. Samples with film thicknesses between 1000 and 60 Å were studied [31, 32]. Figure 9 shows the FIRRAS room temperature absorbance spectra for 1050 and 150 Å thick films of potassium iodide. Easily seen in these spectra is the LO phonon and, on the high-frequency side, a phonon summation band. This sum band is difficult to see in conventional absorption or reflectance spectra of bulk material, but is easily distinguished in FIRRAS spectra.

Our DPM–MPI spectrometer has demonstrated the necessary sensitivity to obtain high-quality far infrared reflection–absorption spectra (FIRRAS) of

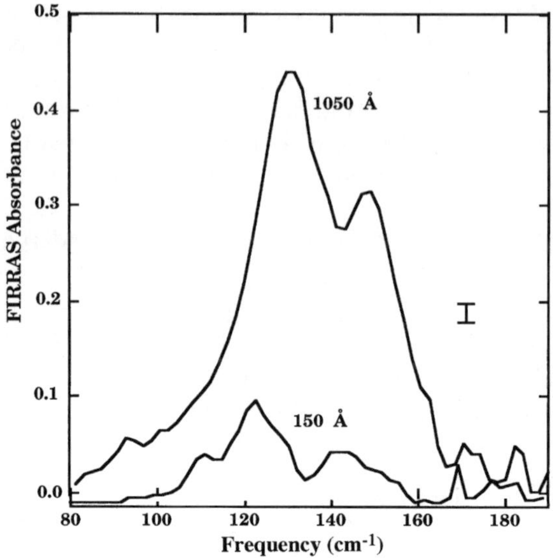

Figure 9 Far-infrared reflectance–absorbance spectra of thin potassium iodide films at room temperature. The error bar provides the approximate noise level of the spectra. The absorbance is as defined by $A(\omega) = 1 - R_p/R_s$. Data from Refs 20 and 21

alkali metal halide films as thin as 50 Å. In addition, the technique has allowed us to explore the morphology of thin films by including the effects of crystallite and void size and shape in the analysis the FIRRAS spectra [32].

7.4 FIRRAS STUDY OF CRYSTALLIZATION OF POLYETHYLENE OXIDE

The study of the vibrational spectra of crystalline polymers is hampered by the inability to produce single-crystal samples of sufficient size for conventional far infrared spectroscopy. However, using thin film sample preparation techniques, it is possible to create samples with high levels of crystallinity. Our study of spun deposited polyethylene oxide (PEO) thin films [37] has demonstrated that the FIRRAS technique allows for the study of polycrystalline polymer samples of sub-micrometer thickness at a higher sensitivity to far-infrared features than can be obtained from transmission spectra of bulk materials. The benefits of FIRRAS in the study of crystalline polymers include the fact that the samples can be prepared with very high levels of crystallinity. Our studies found very high levels of crystallinity for melt recrystallized sub-micrometer PEO films. Figure 10 shows a typical FIRRAS spectrum of a

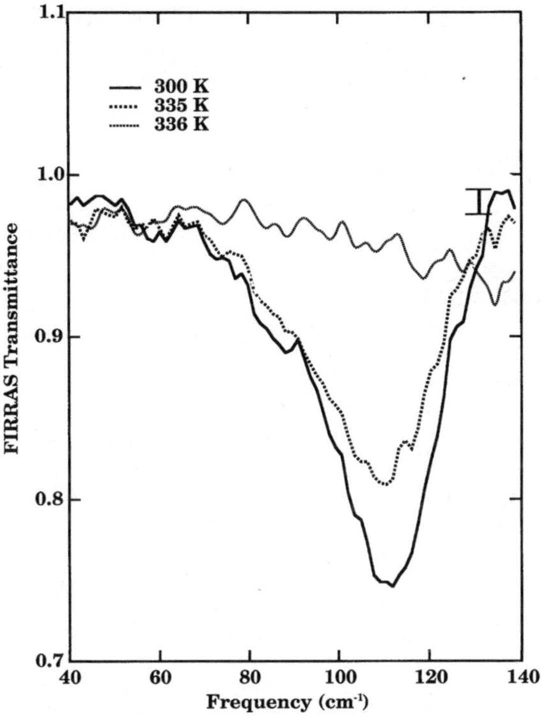

Figure 10 Far-infrared reflectance–absorbance spectra of melt recrystallized polyethylene oxide film with an average thickness of 3900 Å. The melting point of 336.5 K is seen by the loss of the phonon near 110 cm^{-1}. The error bar provides the approximate noise level of the spectra. The transmittance is the ratio of the two polarizations. Data from Ref. 37

3500 Å film of PEO demonstrating the effects of film melting (at 335.5 K) on the far-infrared modes (arising from a loss of helical configuration).

7.5 ADDITIONAL APPLICATIONS

Linear dichroism in polymers has been measured by a simple transmission measurement. We have also explored using the polarization specificity of the DPM–MPI to produce normal incidence reflectance spectra by placing a metal grid polarizer in front of a bulk sample in a near-normal incidence reflectance configuration. Light polarized parallel to the polarizer wires is reflected as from a metallic mirror, while radiation polarized normal to the grid passes through, reflects from the sample and again passes through the grid. We therefore obtain both sample and reference mirror spectra simultaneously and avoid problems of instrument drift and lack of dynamic range in highly reflective spectral regions. This technique could be further extended by evaporating a metallic line grid directly on to the sample. This would avoid many of the problems of reference alignment and positioning.

8 ACKNOWLEDGMENT

I am indebted to my graduate students, present and former, Victor DaCosta, Thomas Fiske, Gregory Spooner, Michael Skolones and Scott Johnson for their many contributions to this research.

9 REFERENCES

1. L.A. Nafie and M. Diem, *Appl. Spectros.* **33**, 130 (1979).
2. L.A. Nafie Polarization modulation FTIR spectroscopy, in *Advances in Applied Fourier Transform Infrared Spectroscopy*, Ed. M.W. Mackenzie, p. 67, Wiley, Chichester (1988).
3. W.G. Golden, D.S. Dunn and J. Overend *J. Catal.*, **71**, 395 (1981).
4. D.H. Martin and E. Puplett, *Infrared Phys.*, **10**, 105 (1970).
5. D.H. Martin, Polarizing (Martin–Puplett) interferometric spectrometers for near- and submillimeter spectra, Ch. 3 in *Infrared and Millimeter Waves*, Ed. K.J. Button, Vol. 6, Academic Press, New York (1982).
6. P.L., Polavarapu, Z. Deng and G.-C. Chen, *Appl. Spectrosc.*, **49**, 229 (1995).
7. P.L. Polavarapu and S. Weibel, *Appl. Spectrosc.*, **50**, 98 (1996).
8. P.L. Polavarapu and S. Weibel, *Appl. Spectrosc.*, **50**, 98 (1996).
9. P.L. Polavarapu and Z. Deng, *Appl. Spectrosc.*, **50**, 686 (11996).
10. P.L. Polavarapu and G.-C. Chen, *Appl. Spectrosco.*, **48**, 1410 (1994).
11. P.R. Griffiths, K.W.V., Every, N. Wright and I.M. Hamadeh, Unconventional applications of FT-IR spectroscopy to the analysis of surfaces, in *Chemical, Biological, and Industrial Applications of Infrared Spectroscopy*, Ed. J.R. Durig, p. 253, Wiley, Chichester (1988).
12. P.A.R. Ade, A.E. Costley, C.T. Cunningham, C.L. Mok, G.F.Neill and T.J. parker, *Infrared Phys.*, **19**, 599 (1979).
13. A.E., Costley, K.H. Hursey, G.F. Neill and J.M. Ward, *J. Opt. Soc. Am.*, **67**, 979 (1977).
14. Analytical Accessories Ltd (SPECAC), Kent, UK.
15. B. Carli and F. Mencaraglia, *Int. J. Infrared MIllimeter Waves*, **2**, 87 (1981).
16. R.C. Jones, *J. Opt. Soc. Am.*, *31*, 23 (1941).
17. W.G. Golden and D.D. Saperstein, *J. Electron Spectrosc.*, **30**, 43 (1983).
18. W.G. Golden, D.D. Saperstein, M.W. Severson and J. Overend, *J. Phys. Chem.*, **88**, 574 (1984).
19 W.G. Golden, Fourier transform infrared reflection–absorption spectroscopy, in *Fourier Transform Infrared Spectroscopy*, Eds P.R. Griffiths and J.A. de Haseth, Vol. 4, pp. 315–345, Academic Press, New York (1985).
20. V.M. DaCosta, Far-infrared spectroscopy of ultra-thin films, *PhD Thesis*, University of California, Davis, CA (1988).
21. V.M. DaCosta and L.B. Coleman, *Appl. Spectrosc.*, **44**, 1301 (1990).
22. M. Skolones and L.B. Coleman, in preparation.

23. M. Skolones, PhD *Thesis*, University of California, Davis, CA (1997).
24. T.G., Fiske, V.M. DaCosta and L.B. Coleman, *Rev. Sci. Instrum.*, **62**, 1861 (1991).
25. D.W. Berreman, *Phys. Rev.*, **130**, 2193 (1963).
26. R.G. Greenler, *J. Chem. Phys.*, **44**, 310 (1966).
27. Y.J. Chabal, *Surf. Sci. Rep.*, **8**, 211 (1988).
28. F.M. Hoffman, *Surf. Sci. Rep.*, **3**, 107 (1983).
29. J.F. Rabolt, F.C. Burns, N.E. Schlotter and J.D. Swalen, *J. Chem. Phys.*, **78**, 946 (1983).
30. J.D. Swalen and J.F. Rabolt, Characterization of orientation and lateral order in thin films by Fourier transform infrared spectroscopy, in *Fourier Transform Infrared Spectroscopy*, Eds P.R. Griffiths and J.A. de Haseth, Vol. 4, pp. 283–314, Academic Press, New York (1985).
31. V.M. DaCosta and L.B. Coleman, *Phys. Rev. B*, **43**, 1903 (1991).
32. T.G. Fiske and L.B. Coleman, *Phys. Rev. B*, **45**, 1414 (1992).
33. G.J.R. Spooner, *PhD Thesis*, University of California, Davis, CA (1994).
34. G.J.R. Spooner and L.B. Coleman, in preparation.
35. A.M. Bradshaw and E. Schweizer, Infrared reflection–absorption spectroscopy of adsorbed molecules, in *Spectroscopy of Surfaces*, Eds R.J.H. Clark and R.E. Hester, p. 413, Wiley, Chichester (1988).
36. K. Ashley and S. Pons, *Chem. Rev.*, **88**, 673 (1988).
37. V.M. DaCosta, T.G. Fiske and L.B. Coleman, *J. Chem. Phys.*, **101**, 2746 (1994).

5

Polarization-Division Interferometry in the Mid- and Far-Infrared Regions

PRASAD L. POLAVARAPU
Department of Chemistry, Vanderbilt University, Nashville, TN 37235, USA

CONTENTS

Principles and Applications of Polarization-Division Interferometry
Edited by P. L. Polavarapu

1 Introduction

As mentioned in the Preface, polarization-division interferometers (PDIs) with different principles of operation are known. In a Wollaston prism-based [1–3] PDI (Figure 1), the incoming light (from an appropriate source or target) is linearly polarized by a polarizer (P) and a Wollaston prism (designated as beamsplitter, BS) divides this polarized light into two components with orthogonal linear polarizations. These two components exiting the prism travel with an angle α (determined from the birefringence of the prism material) between their directions. A linear polarization analyzer (A) placed in their path selects the same polarization component (projected on to the analyzer's axis) for the two rays. An imaging lens (L) focuses these two rays on to a focal plane array detector (D). The signal recorded at each pixel of the detector array represents a different point of the interferogram in spatial domain. The Fourier transform of the interferogram recorded by the array detector yields the spectral information.

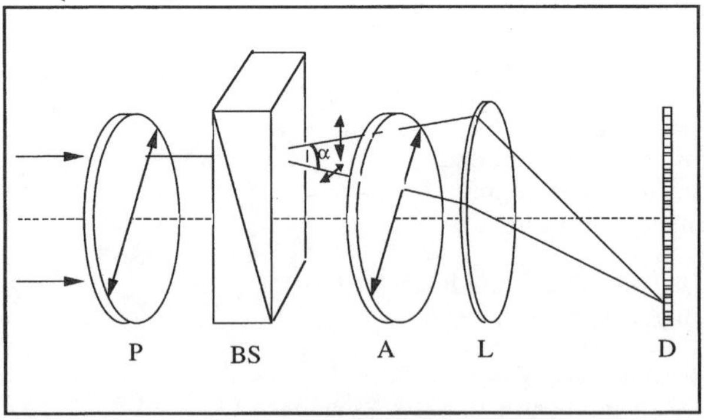

Figure 1 Schematic diagram of a Wollaston prism-based PDI. P, input linear polarizer; BS, polarization-division beamsplitter (Wollaston prism); A, linear polarization analyzer; L, lens; D, array detector

There is no polarization (or any other kind of) modulation involved here, so the Wollaston prism based PDI is a 'static' interferometer and is completely different from the others discussed below. The limitations of this interferometer arise from poor spectral resolution and the limited wavelength region that is accessible with a Wollaston prism, while the advantages result from the stability and compactness of the instrument (as there are no moving parts).

In the approach suggested by Martin and Puplett [4] the input light is linearly polarized and divided into two orthogonal linear polarization components (Figure 2) by a polarization-division beamsplitter (BS) made of wire grids. These two components are brought back to the beamsplitter using two roof-top mirrors M_1 and M_2, which rotate the linear polarizations by 90°. At the zero path difference point (ZPD) of the interferometer, the polarization of the output beam is identical with that of the incoming beam and is defined by the input polarizer P. As the mirror M_2 travels away from the ZPD point the polarization component P_2 undergoes a phase change relative to the polarization component S_1. This phase change is given at a wavenumber $\bar{\nu}_i$ as $2\pi\delta\bar{\nu}_i$ with δ representing the optical path difference. In continuous-scan interferometers $\delta = 2Vt$, where V is the velocity of moving mirror and t is the time; in step-scan interferometers $\delta = 2x$, where x is the displacement of the moving mirror from the ZPD position. It is apparent that the polarization of the output beam depends on the length of the travel by the mirror M_2. For monochromatic input light of wavelength λ, the mirror distances of travel of $\lambda/8$, $2\lambda/8$, $3\lambda/8$, and $4\lambda/8$ correspond, respectively, to phase differences between S_1 and P_2 of $\lambda/4$, $\lambda/2$, $3\lambda/4$ and λ, or equivalently to polarization states of left circular, vertical, right circular and horizontal polarization. As the mirror travels to longer distances this modulation cycle is repeated. For polychromatic light, different wavelengths go through the above mentioned polarization states at different mirror travel positions, since for a

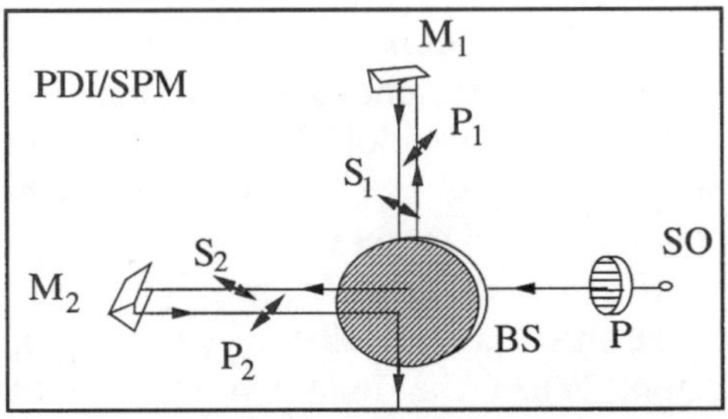

Figure 2 Schematic diagram of a wire-grid beamsplitter-based PDI with single polarization modulation. SO, light source; P, input linear polarizer, BS, polarization-division beamsplitter; P_1, P_2, light components with polarization parallel to the wires of beamsplitter; S_1, S_2, light components with polarization perpendicular to the wires of beamsplitter; M_1, M_2, roof-top mirrors. Configurations for different measurements are shown in Figure 4

longer wavelength correspondingly longer mirror travel is needed to complete one cycle of modulation. Hence this portion of the instrument functions as a polarization modulator with the number of modulation cycles (defined by the length of the mirror travel) being different for different wavelengths.

With the use of a linear polarization analyzer (not shown in Figure 2) placed in the output beam, the polarization modulation discussed above is converted to intensity modulation and the instrument now functions as an interferometer, referred to as a Martin–Puplett interferometer (MPI).

The advantages of this technique (see below) over conventional techniques that use amplitude-division interferometers (ADIs) [5] were not widely recognized [6] in the earlier years (although several applications in the far-infrared and microwave regions have been reported; see Chapters 1–4). A key factor [6] for an increased interest in this interferometry in the last few years is the following: in the absence of a linear polarization analyzer, the detector signal [see Figure 4(d)] contains the *difference* in polarized absorptions (or reflectances) of a sample that is placed in the output beam. An external polarization modulation device would have been needed for such measurements with an ADI; polarization modulation devices are not commonly available for the far-infrared region (owing to the low optical transmission of material-based optical elements), so an MPI provides [6] unique opportunities for circular dichroism measurements in the far-infrared region. The development of this type of instrumentation for the mid-infrared region was undertaken [7] by Bomem Inc., with some modifications to the original design of Martin and Puplett. The rooftop mirrors were replaced by flat mirrors, as a consequence of which the light exiting the interferometer had to be separated from the incoming beam using a small prism. This interferometer was referred to as polarizing Michelson interferometer (PMI), and was commercially marketed for a limited time. Although successful applications were reported [8, 9] with PMI, this design is not expected to provide optimal performance for situations where the quality of polarization states of the light exiting the wire-grid beamsplitter are to be preserved to a high degree until the light passes through the sample. The author's own interest [10-19] in this area developed while evaluating the suggested [6] possibility of measuring far-infrared vibrational circular dichroism.

Owing to the presence of polarization modulation (resulting from the optical path difference) in MPI and PMI, these two wire-grid beamsplitter based instruments can be classified as PDI/SPM, where SPM stands for single polarization modulation. When the light exiting PDI/SPM, encounters an external polarization modulator, such as a photoelastic modulator (PEM), an additional polarization modulation is incorporated. This particular combination is referred [19] to as PDI/DPM, which stands for PDI with double polarization modulation (DPM) (Figure 3). Consider the situation where the stress axis of the PEM and the direction of the wires on the BS are parallel to each other. When the light falling on the PEM has linear polarization at 45° to the PEM's

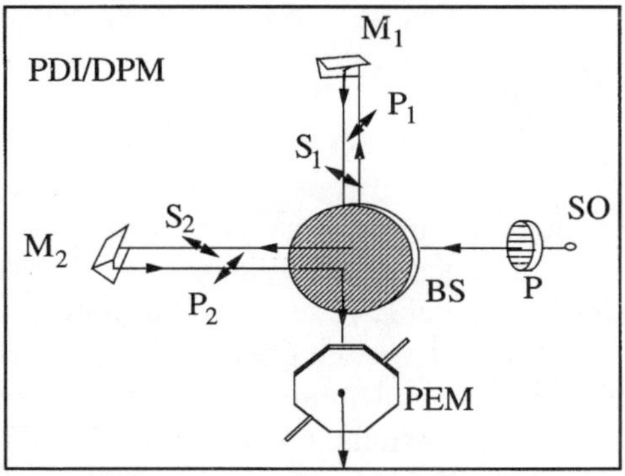

Figure 3 Schematic diagram of a wire-grid beamsplitter-based PDI with double polarization modulation. PEM, photoelastic modulator; other terms and abbreviations as in Figure 2. Configurations for different measurements are shown in Figure 4

axis, the polarization of light exiting the PEM is modulated between two orthogonal circular polarizations (at the characteristic frequency ω_m of the PEM); for circular input polarizations, the output polarizations are modulated between linear polarizations; at the zero crossings of the PEM's sinusoidal modulation cycle, however, the polarization states of the light falling on the PEM are unaffected. As a result, the PEM is adding a higher frequency polarization modulation on top of the slow polarization modulation resulting from the mirror movement. This double polarization modulation has several consequences, as discussed in a later section.

It may be noted that both of the polarization modulations discussed above involve linear and circular polarization states (that is, repeating linear–circular–linear–circular polarization states); these polarization modulations are different from the linear polarization modulations discussed in Chapter 4, where rotating linear polarizers provide the modulation.

For probing the polarization properties of a specimen, PDIs offer definite advantages (see below) over conventional ADIs. The principles and applications of PDI/SPM and PDI/DPM, in the mid- and far-infrared regions, are presented in this chapter.

2 Instrumentation

2.1 Light Sources

A water-cooled globar source is generally used as the light source, but a mercury arc lamp is preferred for the 200–10 cm^{-1} region. Synchrotron radiation (see Chapter 3) serves as a brighter source than the conventional sources.

2.2 Linear polarizers

For the mid-infrared region, linear polarizers with metal wires (spaced at 0.25 μm) deposited on different substrates, such as CaF_2, BaF_2, ZnSe, KBr and KRS-5, are commercially available [20a–c]. Also commercially available are the linear polarizers with metal wires (spaced at ~4 μm) deposited on Mylar film [20d] or a polyethylene substrate [20a], which are suitable for the ~600–10 and 400–10 cm^{-1} regions respectively. The free standing metal wires (spaced at 12 or 25 μm) have frequently been used for the ~400–10 cm^{-1} region, but the above-mentioned polarizers with wires deposited on a mylar substrate are found [14a] to have better polarizing efficiency than the free-standing wire-grid polarizers.

2.3 Beamsplitters

Wire grid-based polarization-division beamsplitters appropriate for the mid-infrared region are not commercially available. A procedure for assembling these beamsplitters was formulated [13a] on a trial and error basis. A Typical procedure involves obtaining two optically matched substrates and depositing the wires on one of these substrates at 0.25 μm spacing. Then these two optical substrates are held together such that the surface containing the wires is sandwiched, and the spacing between the substrates is adjusted using shims. This procedure worked satisfactorily for beamsplitters with a BaF_2 substrate. Owing to the poor infrared transmission of the BaF_2 substrate below 800 cm^{-1}, it became necessary to develop beamsplitters on other substrates that provide long-wavelength transmission. The widely used KBr substrate has infrared transmission up to ~400 cm^{-1}, but the hygroscopic nature of KBr makes the deposition of metal wires on this substrate difficult. To overcome this problem, a thin layer of BaF_2 deposited on KBr substrate was used [13b] as the holding surface for the metal wires. To extend the long-wavelength coverage further, a beamsplitter was also constructed with KRS-5 substrate but the polarization-division quality of this beamsplitter was not satisfactory [13b]. Assemblies of free standing tungsten wires wound on a circular rim have been used as polarization-division beamsplitters for the far-infrared region (see Chapters 1–4). Such devices (5 μm diameter wires at 12 or 25 μm spacing) used in our measurements [14] revealed that the non-uniform spacing between the wires of these beamsplitters is a serious limitation. Metal wires deposited at 4 μm spacing on a thin Mylar film (6 μm thickness) resulted [14a] in a better quality beamsplitter (a matching substrate is not required here). The characteristics of different beamsplitters used in our laboratory are summarized in Table 1.

Table 1 Polarization-division beamsplitters for mid- and far-infrared regions

	Beamsplitter				
	BaF_2	KBr	KRS-5	W wires[a]	Mylar
Support substrate	BaF2	KBr	KRS-5	None	Mylar
Substrate thickness (mm)	3	5	4	–	6×10^{-3}
Spacing between wires (μm)	0.25	0.25	0.25	12	4
Residual %[b]	~1–3[c]	~5[d]	~40[d]	~25[e]	~12[f]

[a] ~5 μm diameter wires.
[b] Residual is defined from the peak positions of $I_r(\delta)$ and $I_b(\delta)$ as $[I_r(\delta)/I_b(\delta)] \times 100\%$ (see Section 3).
[c] Measurements were made with a BaF_2 linear polarizer and ZnSe analyzer.
[d] Measurements were made with a KRS-5 linear polarizer and analyzer.
[e] Measurements were made with a free-standing wire-grid linear polarizer and analyzer.
[f] Measurements were made with a Mylar linear polarizer and KRS-5 analyzer.

2.4 Lenses

Lenses made of CaF_2, BaF_2, ZnSe, and KBr for the mid-infrared region and silicon and polyethylene [21] for the far-infrared region are commercially available. Based on the far-infrared transmission properties, polyethylene lenses are preferred[14] over silicon lenses. Mirrors are commonly used to direct the light exiting the interferometer to the sample and then to the detector. However in applications such as circular dichroism [22] where the quality of the polarization state of exiting light beam needs to be preserved to a high degree, lenses are preferred [22c] over mirrors.

2.5 Detectors

The photoconductive HgCdTe detectors and photovoltaic InSb detectors have been used in our laboratory for the 4000–400 cm^{-1} and 4000–2500 cm^{-1} regions, respectively. For applications in the 400–10 cm^{-1} region, a room temperature DTGS detector (with a polyethylene window) would be a routine choice. However for applications requiring higher sensitivity, liquid helium-cooled bolometers are appropriate (see Chapters 1–4). A silicon bolometer and a hot electron InSb bolometer, with appropriate window material for the region of interest, have been used in our measurements [14].

3 Single polarization modulation

The polarization modulation property of a PDI/SPM, resulting from the moving mirror in the interferometer, can be advantageously used for a variety

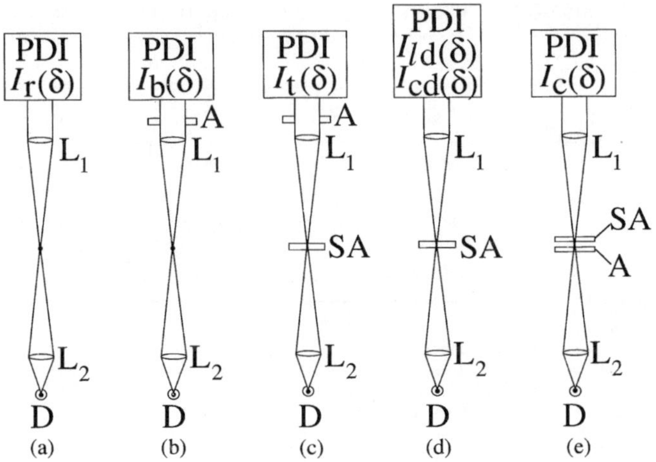

Figure 4 Configurations of PDI for different measurements. L_1, L_2, lenses; A, linear polarization analyzer; D, detector; SA, sample. Configurations (a)–(e) provide, respectively, residual interferogram $I_r(\delta)$, background interferogram $I_b(\delta)$, transmission interferogram $I_t(\delta)$, linear (or circular) dichroism interferogram $I_{ld}(\delta)$ [or $I_{cd}(\delta)$] and birefringence interferogram $I_c(\delta)$; the axis of analyzer A (when present) is at ±45° to that of beamsplitter. To obtain $I_{ld}(\delta)$, the sample's axis is oriented at ±45° to that of the beamsplitter; to obtain $I_c(\delta)$, the sample's axis is oriented parallel (or perpendicular) to that of the beamsplitter

of measurements. Specific configurations suitable for different types of measurements are shown in Figure 4(a)–(e). The interferograms obtained with these configurations and their applications are discussed below.

3.1 Interferograms

The equations (derived [10–13] with the assumption of ideal properties for the optical components involved) applicable for the measurements obtained with configurations shown in Figure 4(a)–(e) are given in Table 2. The notations used are from Ref. 13. Representative examples of the experimentally measured interferograms are shown in Figure 5 and 6.

Residual Interferogram, $I_r(\delta)$

In the configuration shown in Figure 4(a), light exiting the interferometer is transferred by two lenses to the detector, and no other components are placed in the beam. The intensity of light exiting the interferometer, in principle, should not depend on the moving mirror position and therefore the detector signal should not contain any intensity modulation. However a small interferogram

Table 2 Interferograms[a] in PDI with single polarization modulation

Transmission measurements–

$$I_b(\delta) = \int K(\bar{\nu}_i)\cos\theta \, d\bar{\nu}_i$$

$$I_t(\delta) = \int K(\bar{\nu}_i)e^{-a(\bar{\nu}_i)}\cos\theta \, d\bar{\nu}_i$$

$$I_{ld}(\delta) = \int K(\bar{\nu}_i)[e^{-a_s(\bar{\nu}_i)} + e^{-a_p(\bar{\nu}_i)}][a_s(\bar{\nu}_i) - a_p(\bar{\nu}_i)] \times \tfrac{1}{2}\cos\theta \, d\bar{\nu}_i$$

$$I_{cd}(\delta) = \int K(\bar{\nu}_i)[e^{-a_R(\bar{\nu}_i)} + e^{-a_L(\bar{\nu}_i)}][a_L(\bar{\nu}_i) - a_R(\bar{\nu}_i)] \times \tfrac{1}{2}\sin\theta \, d\bar{\nu}_i$$

$$I_c(\delta) = \int K(\bar{\nu}_i)\cos[2\pi\bar{\nu}_i(\delta + d\Delta n)] \, d\bar{\nu}_i$$

Reflectance measurements–

$$I_{db}(\delta) = \int [R_{pb}(\bar{\nu}_i) - R_{sb}(\bar{\nu}_i)]K(\bar{\nu}_i)\cos\theta \, d\bar{\nu}_i$$

$$I_d(\delta) = \int [R_p(\bar{\nu}_i) - R_s(\bar{\nu}_i)]K(\bar{\nu}_i)\cos\theta \, d\bar{\nu}_i$$

$$I_{jb}(\delta) = \int R_{jb}(\bar{\nu}_i)K(\bar{\nu}_i)\cos\theta \, d\bar{\nu}_i; \quad j = \text{s or p}$$

$$I_j(\delta) = \int R_j(\bar{\nu}_i)K(\bar{\nu}_i)\cos\theta \, d\bar{\nu}_i; \quad j = \text{s or p}$$

[a] $K(\bar{\nu}_i) = I^s(\bar{\nu}_i)/4$; $I^s(\bar{\nu}_i)$ is the source intensity at wavenumber $\bar{\nu}_i$; $\theta = 2\pi\delta\bar{\nu}_i$; in continuous-scan interferometers $\delta = 2Vt$, where V is the velocity of moving mirror and t is the time; in step-scan interferometers $\delta = 2x$, where x is the displacement of the moving mirror from the ZPD position; $a(\bar{\nu}_i)$ is the base e absorbance of the sample and is related to decadic absorbance $A(\bar{\nu}_i)$ as $a(\bar{\nu}_i) = 2.303A(\bar{\nu}_i)$; $a_p(\bar{\nu}_i)$ and $a_s(\bar{\nu}_i)$ are the absorbances for parallel and perpendicular linear polarizations; $a_L(\bar{\nu}_i)$ and $a_R(\bar{\nu}_i)$ are the absorbances for left: and right circular polarizations; the differences $[a_s(\bar{\nu}_i) - a_p(\bar{\nu}_i)]$ and $[a_L(\bar{\nu}_i) - a_R(\bar{\nu}_i)]$ are assumed to be small in writing the expressions for $I_{ld}(\delta)$ and $I_{cd}(\delta)$; $R_{pb}(\bar{\nu}_i)$ and $R_{sb}(\bar{\nu}_i)$ are the reflectances of bare surface for p- and s-polarized incident light, respectively; $R_p(\bar{\nu}_i)$ and $R_s(\bar{\nu}_i)$ are the reflectances of the sample (on surface) for p- and s-polarized incident light, respectively.

signal (Figure 5), resulting from the optical imperfections, is generally registered at the detector and this interferogram is referred to as the residual interferogram $I_r(\delta)$.

Background interferogram, $I_b(\delta)$

Intensity modulation can be achieved by placing a linear polarization analyzer (A) in the path of the light exiting the interferometer [see Figure 4(b)]. The detector signal, after eliminating the DC component, represents the instrumental background interferogram $I_b(\delta)$(Figure 5).

Sample transmission interferogram, $I_t(\delta)$

The transmission (or absorption) spectra of samples can be extracted from the interferograms obtained with the configurations shown in Figure 4(b) and 4(c). In the latter configuration, the sample (SA) of interest is placed subsequent to analyzer A and the resulting interferogram is referred to as the sample transmis-

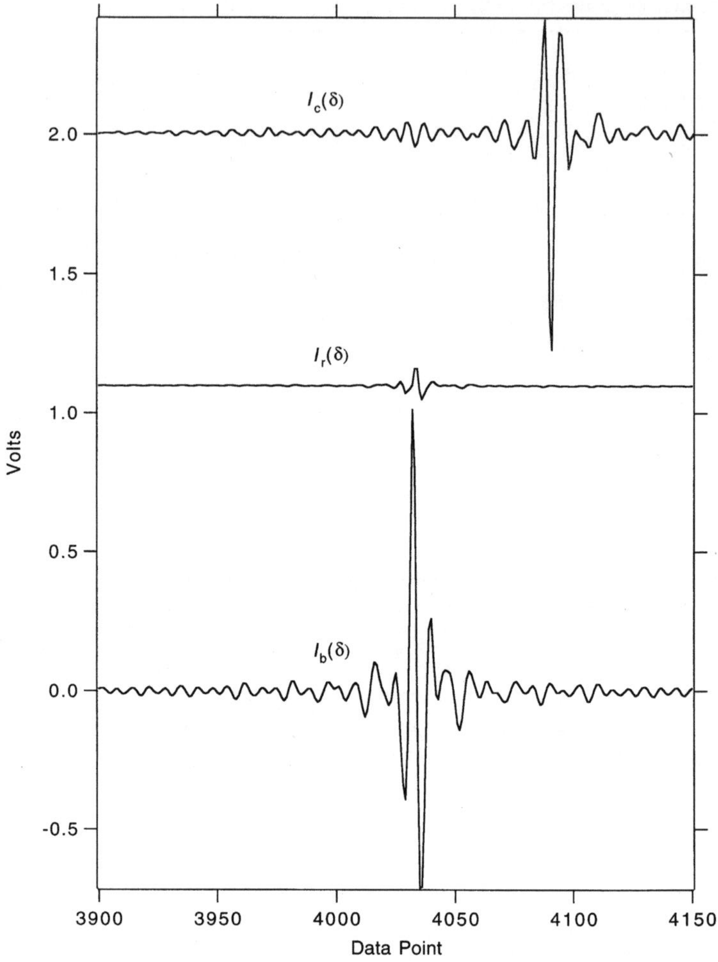

Figure 5 Representative interferograms $I_b(\delta)$, $I_r(\delta)$ and $I_c(\delta)$ obtained with a mid-infrared PDI/SPM using the configurations shown in Figure 4(b), (a) and (e), respectively. To obtain $I_c(\delta)$ a CdSe waveplate was used as the sample. All interferograms are plotted on the same ordinate scale, but are shifted upwards for clarity

sion interferogram $I_t(\delta)$. The logarithm of the ratio of the cosine Fourier transforms of $I_b(\delta)$ and $I_t(\delta)$ yields (see Table 2) the absorption spectrum of the sample. Although the light falling on the sample is linearly polarized, the polarization state of the incident light does not influence the absorption of isotropic samples. However, for oriented samples, the measured absorption depends on the relative orientations of the axes of analyzer and oriented sample. In this case the measured absorption represents polarized absorption.

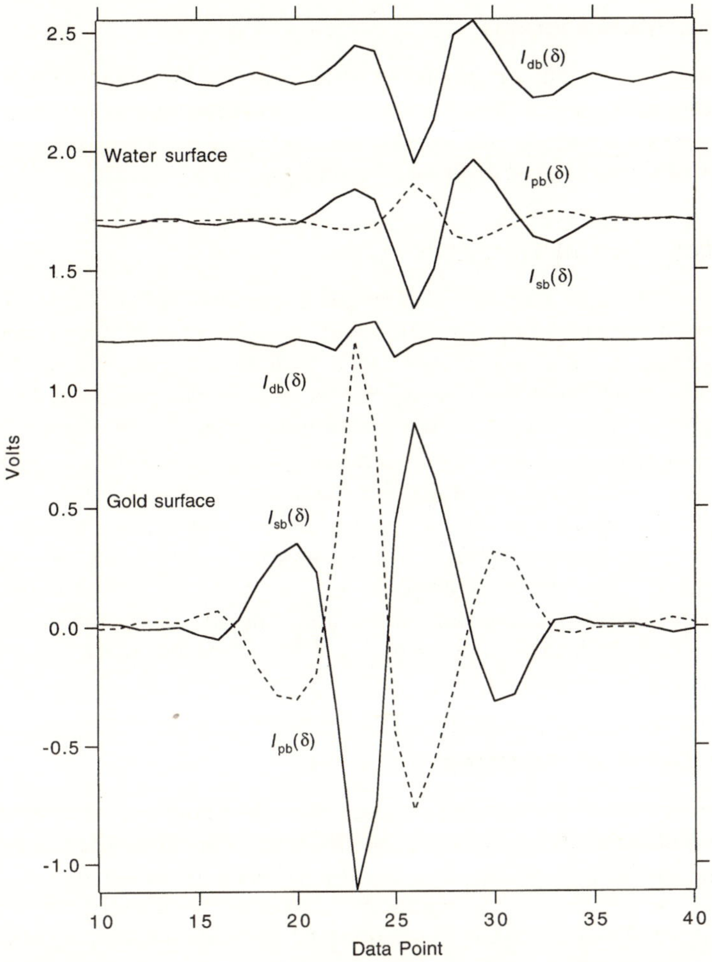

Figure 6 Representative reflectance interferograms obtained with a mid-infrared PDI/SPM. $I_{db}(\delta)$ represents the differential reflectance interferogram of a bare surface [obtained with reflection analog of the configuration shown in Figure 4(a)]. $I_{jb}(\delta)$ represents the polarized reflectance interferogram of a bare surface [obtained with reflection analog of the configuration shown in Figure 4(b)], where j = s for the analyzer's axis perpendicular to the i–r plane and $j = p$ for the axis in the i–r plane. The bottom three traces were obtained for a bare gold-on-mica surface and the top three for a bare water surface. All interferograms are plotted on the same ordinate scale, but some are shifted upwards for clarity. See text for details

Dichroism interferograms

In the configuration shown in Figure 4(d), analyzer A is **not** present and the light exiting PDI passes through the sample to the detector. In this case, the intensity modulation results from the differential polarized absorption properties of the sample as discussed below.

Linear dichroism interferogram, $I_{ld}(\delta)$

The amount of light transmitted through a oriented sample depends on the orientation of the incident polarization with respect to the sample's unique axis. The difference in the absorbance of a sample for two orthogonal linear polarizations of incident light is referred to as linear dichroism (LD). Despite the absence of an analyzer in the configuration shown in Figure 4(d), the intensity modulation occurs when a oriented sample exhibiting LD is present. The interferogram signal depends (see Table 2) on the difference in sample's absorbance for orthogonal linear polarizations; hence this interferogram is referred to as the linear dichroism interferogram, $I_{ld}(\delta)$. For the two orthogonal orientations of the sample (sample's axis at $\pm 45°$ to the direction of the beamsplitter wires), the interferograms are expected to be mirror images of each other [although this property may not be noticeable when the magnitude of $I_{ld}(\delta)$ is less than that of $I_r(\delta)$].

Circular dichroism interferogram, $I_{cd}(\delta)$

For chiral systems in isotropic phase, the absorption of left versus right circularly polarized radiation would be different. This difference is referred to as circular dichroism (CD). Despite the absence of an analyzer, intensity modulation occurs in this configuration due to this circular dichroism property of the sample. Then the interferogram signal depends on the difference in the sample's absorbance for left and right circular polarizations (see Table 2). Hence this interferogram is referred to as the circular dichroism interferogram, $I_{cd}(\delta)$. These interferograms obtained for enantiomers (mirror image systems) are expected to be mirror images [although this feature may not be noticeable when the magnitude of $I_{cd}(\delta)$ is less than that of $I_r(\delta)$].

Birefringence interferogram, $I_c(\delta)$

In the configuration shown in Figure 4(e), both sample and analyzer are present but their relative positions in the optical train are interchanged from those in Figure 4(c). When a specimen exhibiting linear birefringence is used as the sample and the sample's axis is oriented parallel (or perpendicular) to beamsplitter's axis, the peak position of the resulting interferograms will be shifted from those of $I_b(\delta)$ or $I_t(\delta)$ (see Figure 5). The maximum intensity point in the shifted interferogram will appear at an optical path difference of $\delta = \pm d\Delta n$; the two signs here correspond to the two orthogonal (parallel and perpendicular) orientations of the fast axis of the birefringent specimen with respect to the grid direction of the beamsplitter; d is the thickness and Δn is the birefringence (difference between the two refractive indices) of the specimen. Here it is assumed that Δn is independent of the wavelength in the region being probed. The shift in the interferogram peak intensity position can be used for the following purposes: (a) verify the correct function of the interferometer [7, 13, 14] when both d and Δn are known; (b) determine [15] the optical parameter d or Δn, when one of these two parameters is unknown; (c) using a multiple waveplate as the birefringent specimen, the $I_c(\delta)$ obtained can be used for calibrating the linear or circular dichroism signals [7–9, 13].

Reflectance interferograms

For characterizing the species adsorbed on surfaces, the infrared reflectance measurements are more appropriate than the corresponding transmission measurements. In the reflectance measurements, the i–r plane (plane containing the incident ray, reflected ray and surface normal) is used as the reference in stating the polarization state of light. When the electric vector of incident light is perpendicular to the i–r plane it is said to be s-polarized, and, when the electric vector of incident light is parallel to the i–r plane it is p-polarized. Two different configurations are used for reflectance studies. These are similar to the configurations shown in Figure 4(c) and (d) except that the light reflected from the sample surface is collected and focused on to the detector. Polarized reflectance interferograms are obtained in the former arrangement and differential polarized reflectance interferograms are obtained in the latter (see Figure 6). For the bare and sample-deposited surfaces the differential polarized reflectance interferograms are denoted [18] as $I_{db}(\delta)$ and $I_d(\delta)$, respectively. The corresponding polarized reflectance interferograms for the bare and sample deposited surfaces are [18] $I_{jb}(\delta)$ and $I_j(\delta)$, where $j =$ s (for incident polarization perpendicular to the i–r plane) or p (for incident polarization parallel to the i–r plane). The equations representing these interferograms are given in Table 2.

3.2 Instrumental performance

The instrumental efficiency depends on several factors and it is useful to have a set of criteria to compare the relative performances of different instruments. (1) The polarization efficiency is governed by: (a) the polarizing efficiency of the input polarizer; (b) the polarization-division of the input beam by the beamsplitter; and (c) the 90° rotation of the linear polarization by the roof-top mirrors. The extent to which these factors are successfully achieved may be estimated by the relative magnitudes of the residual and background transmission interferograms, $I_t(\delta)$ and $I_b(\delta)$ respectively (see Figure 5). The former should be zero in an ideal instrument. The percentage ratio, defined as $[I_r(\delta)/I_b(\delta)] \times 100\%$, derived from the peak-to-peak magnitudes of these interferograms, provides a measure of the instrumental polarization efficiency. Minimization of this ratio is one criterion that can be used in the optical alignment. The smallest value achieved to date for this ratio, in our laboratory, is 1% (see Table 1). (2) The signal-to-noise ratio (SNR), which depends not only on the optical efficiency but also on the stability of the interferometer, can be gauged by the conventional 100% baseline test. The ratio of two single-beam background spectra, obtained successively, with an 1 min data collection time at 4 cm^{-1} resolution, is used for this purpose. The smallest ratio obtained in the 2000 cm^{-1} region with our mid-infrared instrument is 0.05%. This performance is comparable to that obtained with a similar optical quality ADI. (3) The correct function of the interferometer can be assessed by (a) noting that the background interferograms obtained for the two orthogonal orientations of the output polarization analyzer are expected to be mirror images of each other and (b) verifying the interferogram shifts with a birefringence specimen, as discussed in the previous sub-section.

3.3 Applications

An advantage of PDI over the conventional ADIs arises from its ability to measure the polarization difference signals directly [using the configuration shown in Figure 4(d)]. That is, instead of measuring two large signals and extracting a small difference between them (as would be the case with ADIs), the difference signal is directly registered in PDI, which provides a larger dynamic range. The polarization modulation resulting from the moving mirror in PDI eliminates the need for external polarization modulation devices, which in turn leads to additional advantages [13] over conventional ADIs. (a) In an ADI one half of the source intensity returns to the source and another half is lost at the linear polarizer (used either for obtaining polarized light or for polarization modulation using a PEM). In a PDI, the output polarization anlayzer is not needed for measuring dichroism interferograms [see Figure 4(d)),so only one half of the source intensity is lost (at the input polarizer), [This advantage is not present, however, for polarized absorption measurements, because these measurements with a PDI require an analyzer. For normal absorption measurements PDI is half as efficient as an ADI. This is because a polarizer is not needed for these measurements with an ADI, but an output polarization anlayzer is still needed for measurements with PDI.] (b) The absence of the need for external polarization modulation devices in a PDI makes it feasible to undertake micro- to nanosecond temporal polarization difference studies [17]. On the other hand, temporal polarization difference studies on the microsecond (or faster) time-scale are not as convenient with ADIs because the operating frequency of PEM in an ADI overlaps with this time domain.

Hence the nature of signals present in PDI makes it an ideal method for measuring the (both static and time resolved) polarization properties of the sample, some such properties being linear dichroism and birefringence of oriented samples and circular dichroism of chiral molecules in isotropic phase. Representative applications in these areas are presented in this section.

Polarized absorption and linear dichroism spectra

Liquid crystals can be assembled with a specific orientation. When incorporated into a cell with the inner surfaces coated with polyimide [or poly(vinyl alcohol)], liquid crystals attain a well defined orientation. A liquid crystal cell therefore exhibits birefringence and can be used as a retarder. For linearly polarized light with the polarization axis at 45° to the oriented axis of liquid crystal (see Figure 7), the incident polarization can be resolved into two components one parallel to the oriented axis and the other perpendicular to that axis. These two components have different velocities when transmitted through the liquid crystal owing to the birefringence of the liquid crystal. The nature of the polarization of light exiting the liquid crystal depends on the magnitude of its birefringence. In the presence of an electric field across the liquid crystal cell, the orientation of liquid crystal molecules tilts towards out-of-the plane of the cell, so the effective birefringence of the cell changes. This effective birefringence depends on the applied voltage. Then, by modulating the magnitude of voltage applied to the liquid crystal, the polarization of light exiting the liquid crystal is also modulated. This is the principle behind the

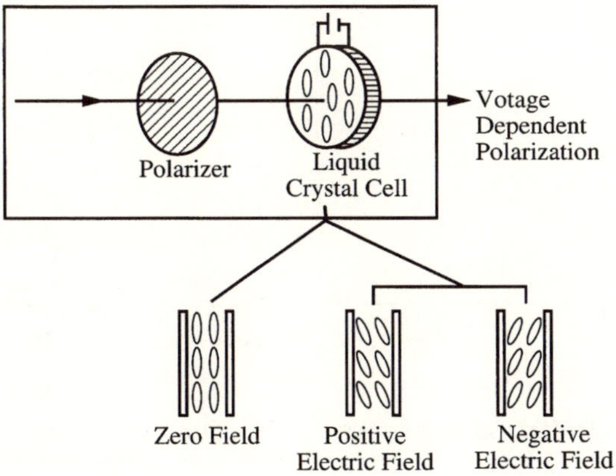

Figure 7 Depiction of the influence of electric field on the orientation of liquid crystal molecules

development of liquid crystal variable retarders [23]. In order to prevent the run-off of liquid crystal molecules, a bipolar square-wave voltage, at a frequency (2 kHz) much greater than the desired frequency of polarization modulation (~10 Hz), is needed. The bipolar voltage makes the orientation of liquid crystal molecules shift from in-plane to above and below the plane of the cell as the polarity changes from positive to negative, but since the magnitude

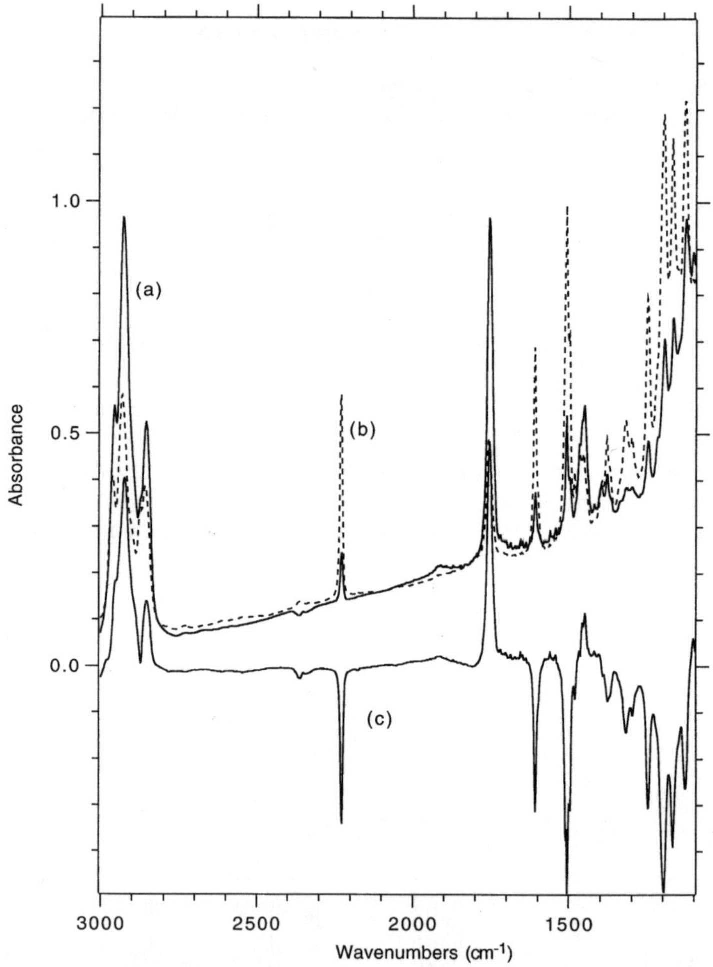

Figure 8 (a, b) Polarized absorption spectra and (c) linear dichroism spectrum of ZLI-2140 liquid crystal cell with anti-reflection coated germanium windows. The linear dichroism spectrum is derived as the static difference between the polarized absorption spectra, which were obtained with the configuration shown in Figure 4(c)

of applied voltage is the same for both polarities the effective birefringence remains constant in both halves of the bipolar square wave. When the magnitude of this bipolar electric field is changed, the polarization state of the light exiting the liquid crystal changes. Hence, liquid crystals can be used as inexpensive polarization modulators, as long as the modulation speeds of ~10 Hz or slower are adequate. The electric field-induced changes in the

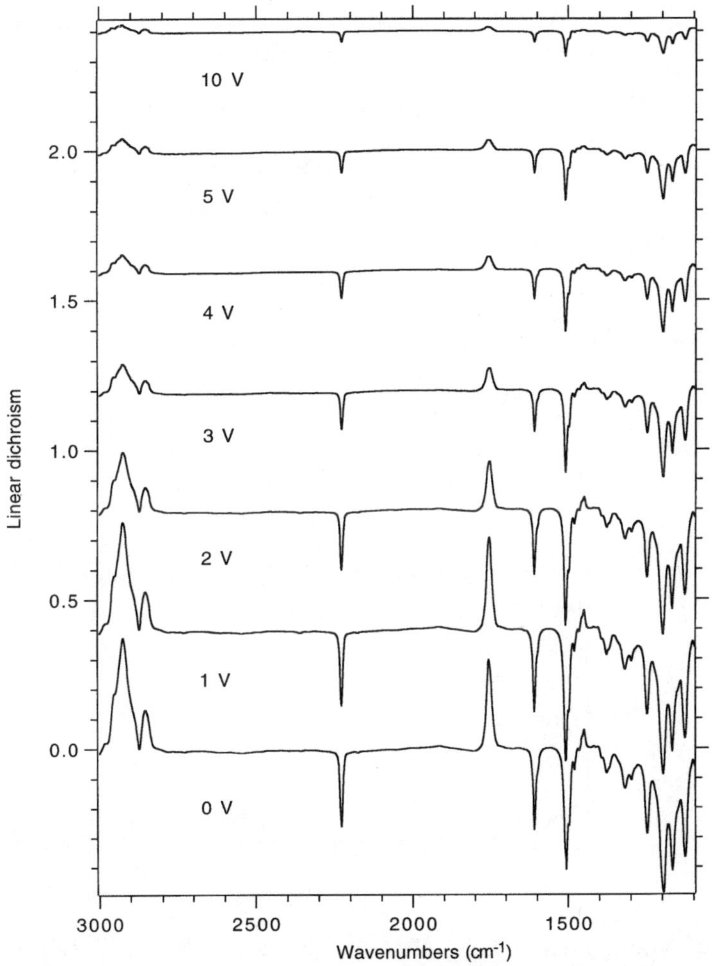

Figure 9 Linear dichroism spectra of ZLI-2140 liquid crystal cell obtained with a mid-infrared PDI/SPM at different applied voltages. These spectra were derived directly from the normalized cosine Fourier transform of $I_{\mathrm{ld}}(\delta)$ obtained with the configuration shown in Figure 4(d). The ordinate scale is the same for all spectra, but the spectra are shifted upwards for clarity

orientation of the liquid crystal molecules can be probed by monitoring the linear dichroism and birefringence properties as discussed below.

The experimental data obtained for a ZLI-2140 liquid crystal [23] (a mixture of seven compounds [23b] with CN, CO, benzyl and cyclohexyl functional groups), with PDI are shown in Figure 8–10. The liquid crystal was aligned

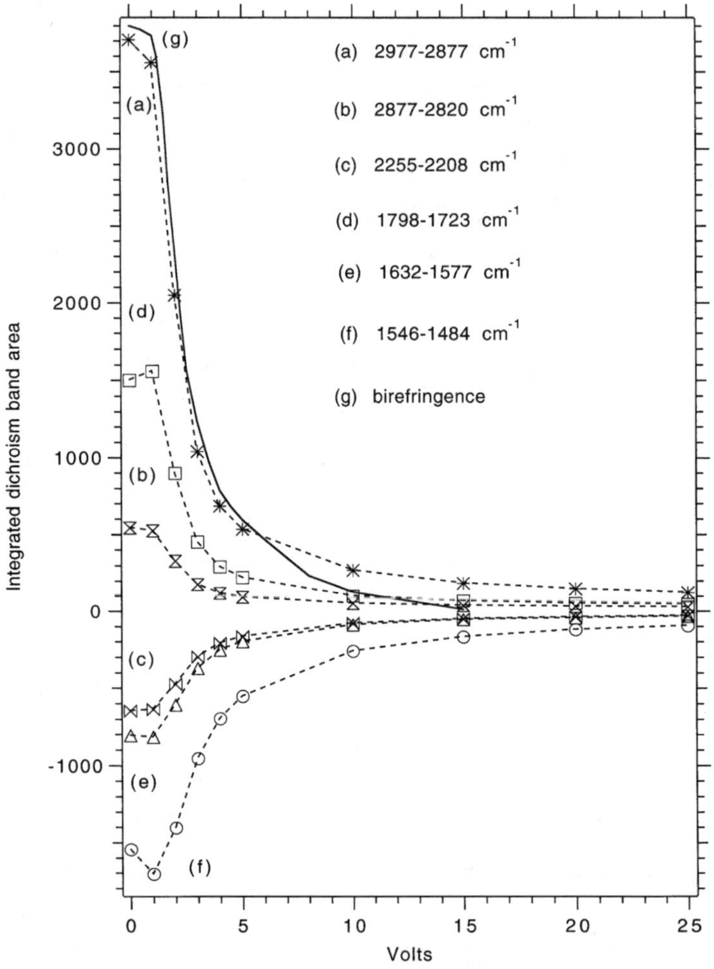

Figure 10 Voltage-dependent changes in the areas of selected linear dichroism bands of ZLI-2140 liquid crystal. These areas (in arbitrary units) were derived from the spectra shown in Figure 9. The listed ranges are the band spreads used in deriving the areas. Trace (g) represents the voltage-dependent birefringence [derived from $I_c(\delta)$]. All band areas are plotted on the same ordinate scale, but the birefringence values (see Figure 11) are scaled by a constant to display in the same plot

and held between anti-reflection coated germanium windows. The absorption spectrum obtained with configuration shown in Figure 4(c) is equivalent to the polarized infrared spectrum obtained with conventional FTIR spectrometers. Two polarized absorption spectra (with analyzer's axis parallel and perpendicular to sample's axis and no voltage applied to the liquid crystal) are shown in Figure 8(a) and 8(b). The difference in the absorbance spectra obtained with two orthogonal orientations of the analyzer's axis, while keeping the sample orientation fixed, represents the linear dichroism obtained as the static difference. This difference is shown in Figure 8. The normalized cosine Fourier transform of $I_{ld}(\delta)$ [obtained with the configuration shown in Figure 4(d) should give the same result (see Table 2) as that obtained in the above-mentioned static difference method, for the same sample orientation, but with better signal quality. The dichroism spectrum obtained in this manner (which can be referred to as dynamic difference measurement) is shown as the bottom trace in Figure 9. These dynamic spectra obtained as a function of the voltage applied to the liquid crystal are also shown in Figure 9. As expected, the dichroism band intensities decrease as the magnitude of voltage applied increases. The variations in integrated dichroism band areas, for selected vibrational bands, as a function of applied voltage are shown in Figure 10. A detailed discussion of the changes seen in band areas is beyond the scope of this chapter, but it is useful to note that the pattern of changes seen between 0 and 1 V is different from that seen between 1 and 5 V, indicating differential molecular orientational response as a function of applied voltage.

Birefringence measurements

A liquid crystal cell, assembled with anti-reflection coated germanium windows and placed before the analyzer [as shown in Figure 4(e), is oriented such that the liquid crystal axis is parallel to the grid direction of the beamsplitter. The shifts in the interferogram peak position, observed as a function of the voltage applied to the liquid crystal, are used to derive [15] the birefringence as discussed earlier. For a better accuracy in determining these shifts, the central portion of interferograms were fitted to a gaussian curve and the maximum intensity position was derived through curve fitting. The observed changes in birefringence of a ZLI-2140 liquid crystal and of a 4′-pentyl-4-biphenylcarbonitrile liquid crystal are shown in Figure 11 as a function of voltage. The birefringence data obtained for a ZLI-2140 liquid crystal is also included in Figure 10 to compare the pattern of changes in birefringence with those in dichroism band areas.

Reflectance measurements

The polarization dependence of infrared reflection from thin films on surfaces has been the basis for the widespread use of infrared spectroscopy in surface

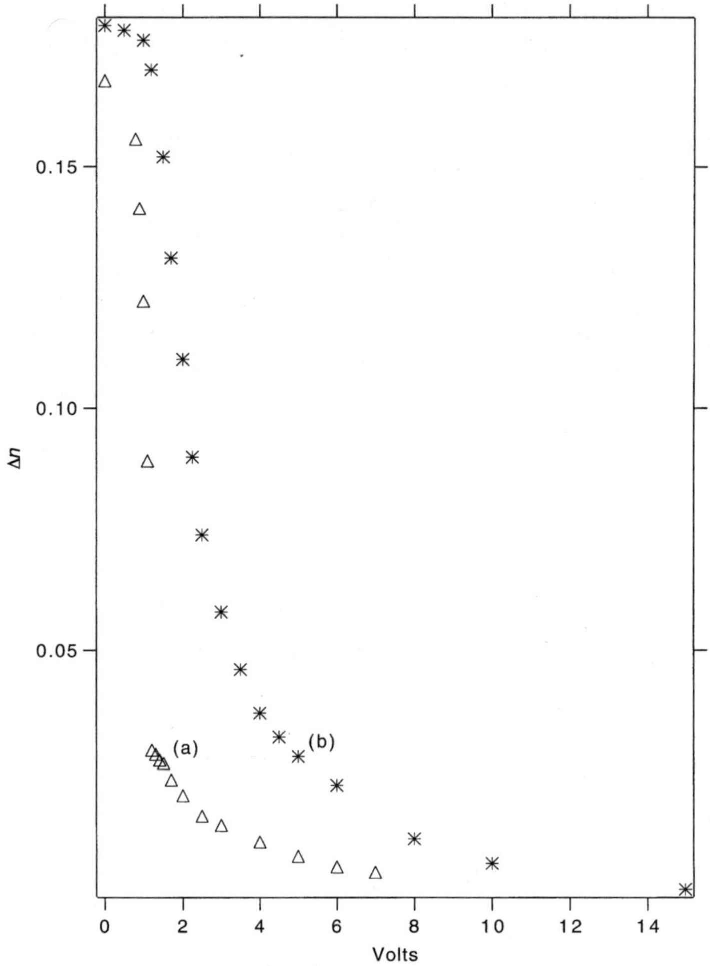

Figure 11 Mid-infrared birefringence of two different liquid crystal cells determined from $I_c(\delta)$. (a) 4′-Pentyl-4-biphenylcarbonitrile liquid crystal; (b) ZLI-2140 liquid crystal. The latter data were also included in Figure 10

characterization. The applications of PDI for samples at air–water [18] and air–metal interfaces [8, 18] are presented in this section. The differential and polarized reflectance interferograms obtained for a bare water surface are shown in Figure 6 (top half). The differential interferogram $I_{db}(\delta)$ has a fairly large magnitude because of the large difference in the polarized reflectances. Owing to the intrinsic contribution from the water surface, the normalized cosine Fourier transform of $I_{db}(\delta)$ yields an S-shaped intensity distribution. This background distribution determines the shape of the spectrum obtained for the sample on the water surface and subtraction of the bare water surface spectrum from that obtained with the sample on the water surface is necessary to extract the sample spectrum. The subtracted spectrum is shown in Figure 12 for a monolayer of arachidic acid on water.

The differential and polarized reflectance interferograms obtained for a bare gold-on-mica surface are included in Figure 6 (bottom half). Owing to the similar magnitudes of polarized reflectances, the differential interferogram $I_{db}(\delta)$ has a very small magnitude. The differential polarized reflectance spectrum of poly(vinyl acetate) obtained after subtracting the corresponding spectrum of blank gold surface is shown in Figure 13.

Circular dichroism spectra

The linear dichroism signals (discussed earlier for oriented molecules) are usually of the same magnitude as the corresponding absorption magnitudes. However, the circular dichroism signals, exhibited by chiral molecules in the isotropic phase, in the mid- and far-infrared regions are expected to be four to five orders of magnitudes smaller than the corresponding absorption magnitudes. The signal quality obtainable in the far-infrared region is usually limited

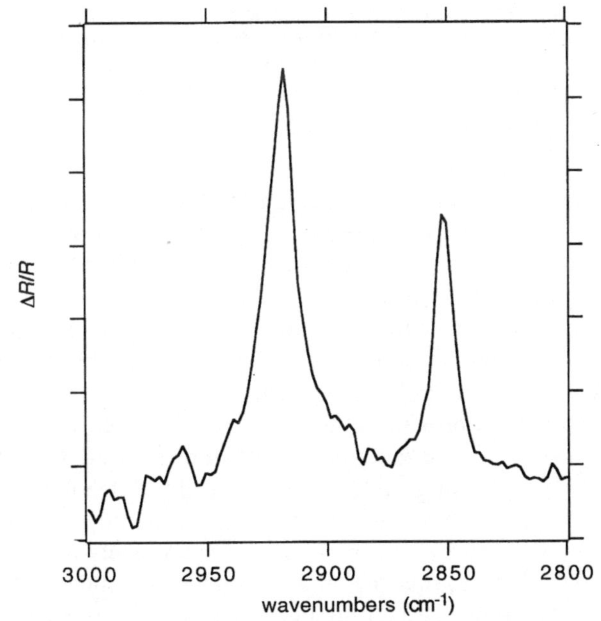

Figure 12 Normalized differential reflectance spectrum cf arachidic acid monolayer on a water surface obtained with PDI/SPM, after subtracting the corresponding spectrum of the blank water surface

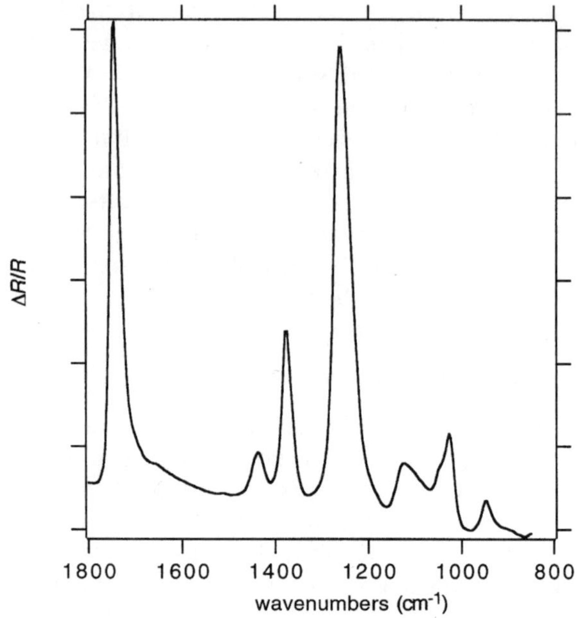

Figure 13 Normalized differential reflectance spectrum of poly(vinyl acetate) on a gold surface obtained with PDI/SPM, after subtracting the corresponding spectrum of the blank gold surface. The thickness of the poly(vinyl acetate) film (deposited on the gold surface) was estimated to be 200 nm. Data were taken from Ref. 18 and replotted

because of the low emissivity, as expected for blackbody radiation, from the conventional light sources. This problem is further compounded by the poor far-infrared transmission of the material-based external polarization modulation devices. The measurement of circular dichroism in the far-infrared region has not yet been achieved with conventional techniques. The material-based external polarization modulation devices, however, are not required for dichroism measurements with PDI, so PDI provides an efficient approach for measuring the circular dichroism in the far-infrared region [14].

Circular dichroism measurements in the mid-infrared region have been successfully made [22] using ADIs, where the desired polarization modulation is achieved with a PEM. The ability for measuring circular dichroism with PDI can be assessed by making these measurements in the mid-infrared region with PDI and comparing the results with those obtained using conventional ADIs. Such measurements have been demonstrated [9, 13]. A representative spectrum is shown in Figure 14. Following the success with these mid-infrared circular dichroism measurements, attempts were made to measure the circular dichroism measurements in the far-infrared region.

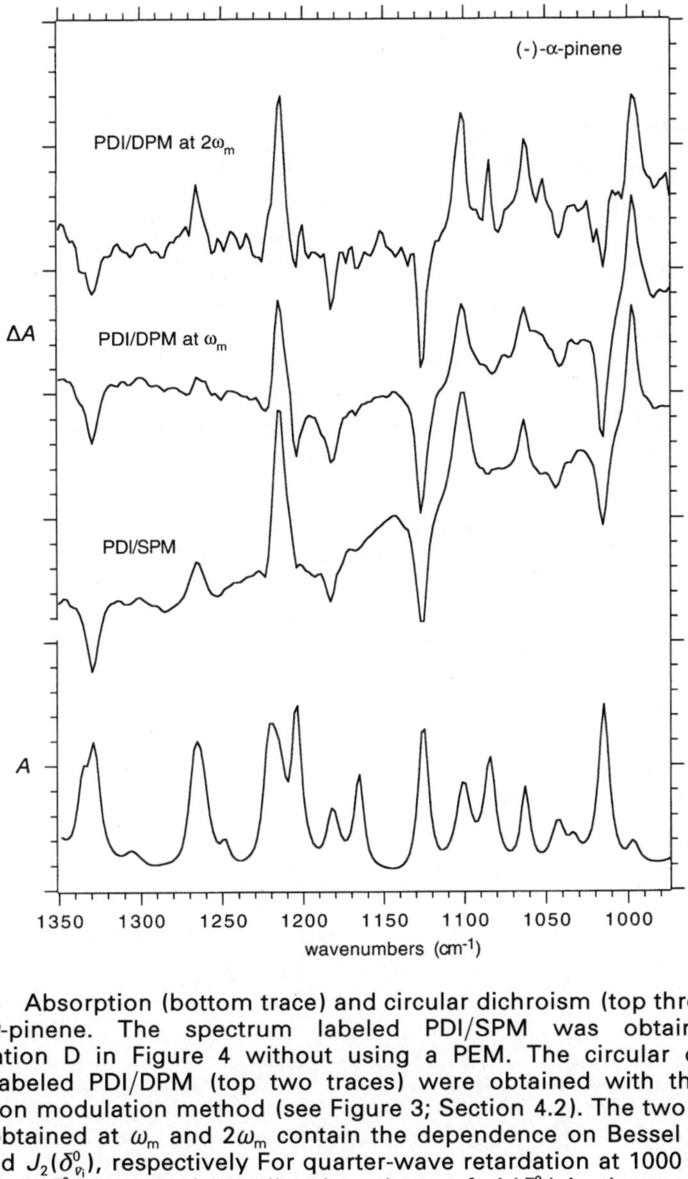

Figure 14 Absorption (bottom trace) and circular dichroism (top three traces) of (−)-α-pinene. The spectrum labeled PDI/SPM was obtained with configuration D in Figure 4 without using a PEM. The circular dichroism spectra labeled PDI/DPM (top two traces) were obtained with the double polarization modulation method (see Figure 3; Section 4.2). The two PDI/DPM spectra obtained at ω_m and $2\omega_m$ contain the dependence on Bessel functions $J_1(\delta^0_{\bar{\nu}_i})$ and $J_2(\delta^0_{\bar{\nu}_i})$, respectively For quarter-wave retardation at 1000 cm^{-1}, the values of $J_2(\delta^0_{\bar{\nu}_i})$ are much smaller than those of $J_1(\delta^0_{\bar{\nu}_i})$ in the wavenumber range shown, and therefore circular dichroism measured at $2\omega_m$ has more noise than that measured at ω_m

These measurements in the far-infrared are not yet routine and further improvements are needed [14].

Time-resolved dichroism measurements

In conventional methods using ADIs, time-resolved linear dichroism can be extracted as the difference between two separately measured time-resolved polarized absorptions. Such measurements suffer from potential instrumental instabilities. When the polarization is modulated using a PEM, the time-scale of polarization modulation by PEM (usually 30–100 KHz) overlaps with the microsecond time-scale, where the temporal behavior of a sample is being studied. For this reason, microsecond time-resolved linear dichroism measurements have not been performed to date using conventional interferometers that use a PEM for polarization modulation. Since PDI measures the difference signal directly [see the configuration in Figure [4(d)], time-resolved dichroism can be measured [17] with PDI fairly conveniently. The experimental configuration to realize direct time-resolved dichroism measurements is depicted in Figure [15(a)]. The analog detector signal is processed by a boxcar integrator with selectable time delays and gate widths. A reference signal, derived from a controller that supplies stimulus to initiate the temporal changes in the sample, is used to define the delay (from the time stimulus given) T_d,

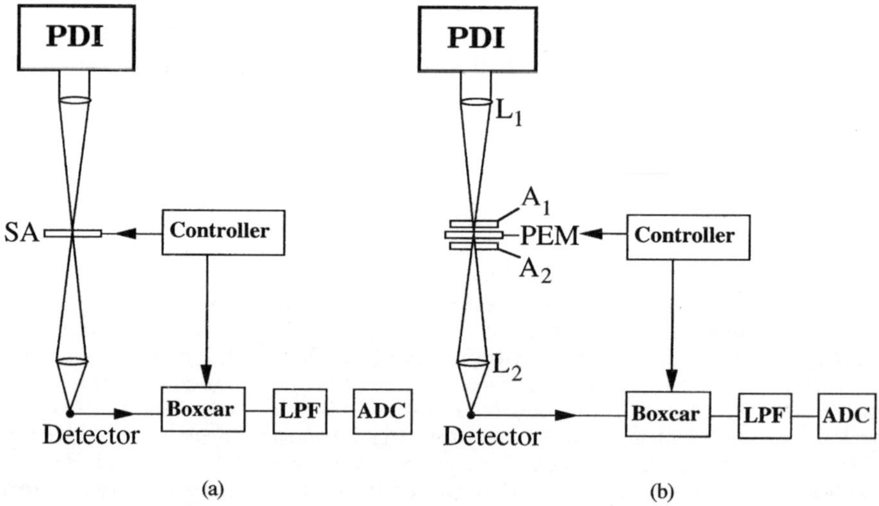

Figure 15 Schematic diagram of the experimental set-up for time-resolved dichroism measurements with PDI/SPM. LPF, low-pass filter; ADC, analog-to-digital converter. The configuration in (a) is for a general sample and that in (b) is for a test case with an analyzer–PEM–analyzer combination serving as a test 'sample'

and gate width (time width during which the detector signal is allowed to reach the digitizer) T_g. The former is chosen based on the time resolution needed and the latter on both time resolution and signal-to-noise ratio. In continuous-scan interferometers, the asynchronous perturbation method [24] can be used for signal digitization. The asynchronous perturbation method requires that the external stimulation frequency f_e be greater than $\sim 4V\nu_{max}$, where V is the velocity of moving mirror and ν_{max} is the maximum wavenumber in the detected spectral region. For $V = 0.6$ cm^{-1}/s and $\nu_{max} = 8000$ cm^{-1}, f_e should be greater than $4 \times 0.6 \times 8000 = 19.2$ kHz.

A nice demonstration of these concepts can be made [17] using analyzer–PEM–analyzer combination as the 'sample' [see Figure 15(b)] and monitoring the temporal nature of the PEM function. When an analyzer is placed in the path of the light exiting PDI [Figure 4(b)], the function of PDI is identical with that of a conventional ADI providing transmission interferograms. Subsequent to this analyzer, a PEM and a second analyzer were placed with the polarization axes of the two analyzers orthogonal to each other and the axis of the PEM parallel to the grid direction of the beamsplitter (at 45° to the analyzers' axes). The PEM modulated the linear polarizations at 75 kHz, which is greater than the above-mentioned 19.2 kHz. The detector signal was analyzed with a boxcar averager using a gate width of 0.5 μs. The transmission interferograms were recorded with time delays at 1 μs intervals and Fourier transformed to obtain time-resolved spectra. In the present experiment, the voltage applied to PEM was chosen to give one half-wave retardation at 2000 cm^{-1}. Then the intensity at 2000 cm^{-1} is expected to change [17] as

$$I(\bar{\nu}_i) \simeq \tilde{I}^s(\bar{\nu}_i)\cos\delta^m_{\bar{\nu}_i} \simeq \tilde{I}^s(\bar{\nu}_i)[J_0(\delta^0_{\bar{\nu}_i}) + 2J_2(\delta^0_{\bar{\nu}_i})\cos 2\omega_m t + 2J_4(\delta^0_{\bar{\nu}_i})\cos 4\omega_m t + \ldots]$$

where $I^S(\bar{\nu}_i)$ is the source intensity, $\delta^m_{\bar{\nu}_i} = \delta^0_{\bar{\nu}_i}\sin t\omega_m t$ with $\delta^0_{\bar{\nu}_i}$ representing the retardation introduced by PEM at time zero and $\omega_m = 2\pi\nu_m$ is the frequency of stress voltage applied to PEM (37.5 kHz in the present case); $J_0(\delta^0_{\bar{\nu}_i})$, $J_1(\delta^0_{\bar{\nu}_i})$ and $J_2(\delta^0_{\bar{\nu}_i})$ are the Bessel functions. For the wavenumber where PEM imparts half-wave retardation, higher order terms involving $J_6(\delta^0_{\bar{\nu}_i})$, etc., are negligible. The simulated intensity pattern resulting from the above equation is displayed in Figure 16 (lower half) and the observed time-resolved intensity pattern is displayed in figure 16 (upper half). The observed intensity at 2000 cm^{-1} was in fact modulated with a period of 14 μs, which is in agreement with the expected value of 13.3 μs, noting that the time resolution employed was 1 μs and integrating time was 0.5 μs. These data demonstrate the capability of PDI for time-resolved dichroism measurements. The temporal behavior of a liquid crystal was investigated [17] in this manner.

The above-described procedure for microsecond time-resolved linear dichroism can also be extended to circular dichroism. The temporal evolution

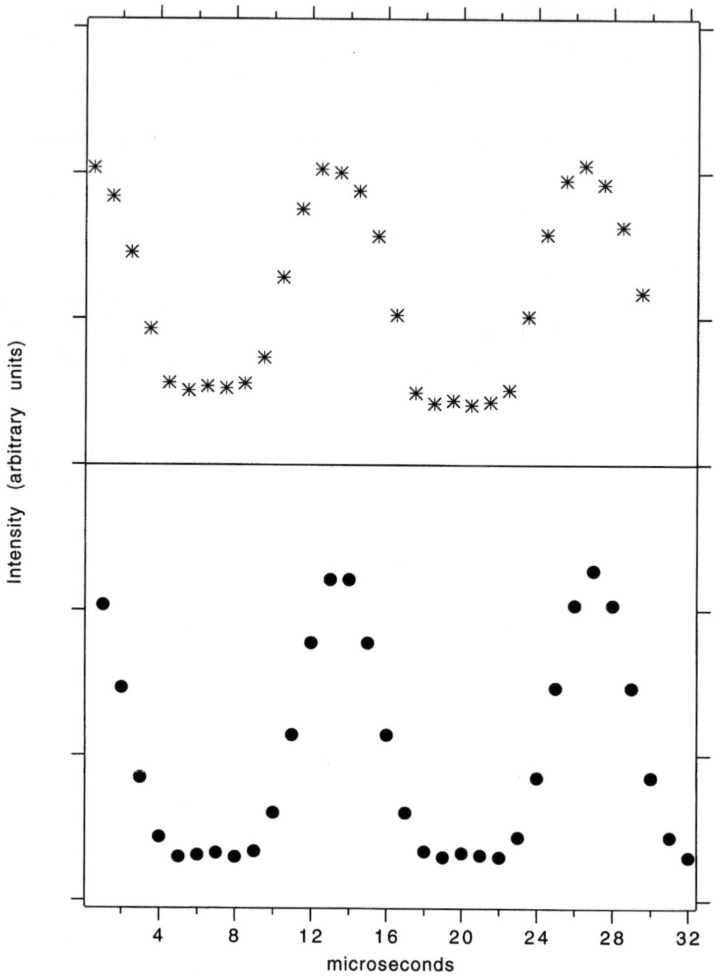

Figure 16 Microsecond time-resolved transmission measured with the configuration shown in Figure 15(b). The PEM controller was set to introduce half-wave retardation at 2000 cm^{-1}. The top and bottom panels display the experimental data and theoretical predictions, respectively. Data were taken from Ref. 17 and replotted

of differential polarized reflectance at interfaces can also be probed similarly. For time-resolved birefringence, the sample will be placed before the analyzer [as in Figure 4(e)] and the shifts in interferogram peak positions will be monitored for different T_d values. The measurements mentioned in this paragraph are yet to be undertaken.

Infrared dichroic imaging

As the configuration shown in Figure 4(d) permits the measurement of polarized absorption difference signals directly, further opportunities that would not have been available with conventional ADIs can be realized. Since isotropic achiral solvent samples such as water exhibit neither linear dichroism nor circular dichroism, such samples are expected to be 'transparent' for measurements with the configuration in Figure 4(d). The advantages of measuring linear dichroism with PDI in the presence of interfering water absorption have been demonstrated [16]. These preliminary studies indicate that infrared dichroic imaging with PDI is potentially important and worth pursuing. The dichroic images generated from the linear dichroism spectral property are expected to reveal the spatial distribution of oriented solutes. For example, polymer flow in water solvent or a biomembrane surface in the presence of a water environment are some examples that are expected to benefit from dichroic imaging studies, which are yet to be undertaken.

4 Double polarization modulation

Here we consider PDI/DPM [19] where one polarization modulation results from the moving mirror in the interferometer and an additional polarization modulation is incorporated using a PEM (see Figure 3). Different configurations shown in Figure 4 are applicable here also, except that the signal processing would be different from that in PDI/SPM. In a PDI/DPM the detector signal contains harmonics at ω_m, $2\omega_m$, etc., because of the oscillatory function of PEM at ω_m. Note that the residual interferograms however are expected to be similar in PDI/DPM and PDI/SPM. The equations [19] applicable for some measurements of interest are given in Table 3.

4.1 Polarization modulation analysis

The cosine Fourier transform of $I_b(\delta)$ obtained with PDI/DPM [after eliminating the signals at $n\omega_m$ ($n = 1$, 2 etc.); see Table 3] gives the source intensity distribution multiplied with the Bessel function $J_0(\delta^0_{\bar{\nu}_i})$. In the background spectrum obtained with a PDI/SPM (see Table 2) this Bessel function would not be involved, so the ratio of the cosine Fourier transform of $I_b(\delta)$ obtained with PDI/DPM to that obtained with PDI/SPM yields the $J_0(\delta^0_{\bar{\nu}_i})$ spectrum. The value of $J_0(\delta^0_{\bar{\nu}_i})$ at a given wavelength depends on the retardation $\delta^0_{\bar{\nu}_i}$ imparted by the PEM at that wavelength. The PEM controller was set to provide half-wave retardation at 1, 2, 3, 4 or 5 μm and the

Table 3 Interferograms[a] in PDI with double polarization modulation

$$I_b(\delta) = \int K(\bar{\nu}_i)[J_{\text{even}} \cos\theta - J_{\text{odd}} \sin\theta] \, d\bar{\nu}_i$$

$$I_t(\delta) = \int K(\bar{\nu}_i) e^{-a(\bar{\nu}_i)} [J_{\text{even}} \cos\theta - J_{\text{odd}} \sin\theta] \, d\bar{\nu}_i$$

$$I_{ld}(\delta) = \int K(\bar{\nu}_i)[e^{-a_s(\bar{\nu}_i)} + e^{-a_p(\bar{\nu}_i)}][a_s(\bar{\nu}_i) - a_p(\bar{\nu}_i)] \times \tfrac{1}{2}[J_{\text{even}} \cos\theta - J_{\text{odd}} \sin\theta] \, d\bar{\nu}_i$$

$$I_{cd}(\delta) = \int K(\bar{\nu}_i)[e^{-a_R(\bar{\nu}_i)} + e^{-a_L(\bar{\nu}_i)}][a_L(\bar{\nu}_i) - a_R(\bar{\nu}_i)] \times \tfrac{1}{2}[J_{\text{even}} \sin\theta - J_{\text{odd}} \cos\theta] \, d\bar{\nu}_i$$

[a] For definitions of terms see Table 2;

$$J_{\text{even}} = J_0(\delta^0_{\nu_i}) + 2 \sum_{n=1}^{\infty} J_{2n}(\delta^0_{\nu_i}) \cos 2n\omega_m t;$$

$$J_{\text{odd}} = 2 \sum_{n=1}^{\infty} J_{2n-1}(\delta^0_{\nu_i}) \sin[(2n-1)\omega_m t];$$

$\omega_m = 2\pi\nu_m$ is the operating frequency of PEM.

experimental $J_0(\delta^0_{\bar{\nu}_i})$ spectra obtained are shown in Figure 17 (as dashed traces). The predicted $J_0(\delta^0_{\bar{\nu}_i})$ spectra for the same retardations are also displayed in Figure 17 (as solid traces). A comparison of the two sets of spectra indicates that PDI can be used as a *polarization modulation analyzer*; that is, polarization modulation devices can be evaluated for the actual retardation imparted by them.

4.2 Dichroism measurements

In a PDI/DPM the linear dichroism signal $\Delta a(\bar{\nu}_i)$ appears (see Table 3) at all integer multiples of ω_m (i.e. at $n\omega_m$, with $n = 0, 1, 2$, etc.), so one may extract the linear dichroism components at non-zero $n\omega_m$ ($n = 1, 2$ etc.) with the use of a lock-in amplifier (tuned to ω_m, ω_m, etc.) and another component ($n = 0$) without the use of a lock-in amplifier. The same is true with the circular dichroism signal. For one-half or one-quarter wave retardation in the mid-infrared region, the components of $n\omega_m$ with $n > 2$ are not important because the magnitudes of the corresponding Bessel functions $J_n(\delta^0_{\bar{\nu}_i})$ are small. The individual $n\omega_m$ ($n = 0$, 1 and 2) components of dichroism (linear or circular dichroism) can be measured [19] (see Figure 14) to verify the underlining principles of PDI/DPM. However, if all three $n\omega_m$ ($n = 0$, 1 and 2) components can be measured simultaneously (which can be achieved with digital signal processing techniques), then the total linear or circular dichroism signal measured with PDI/DPM would be about two times larger [19] than that measured with a PDI/SPM. Since PDI/SPM itself is two times more efficient

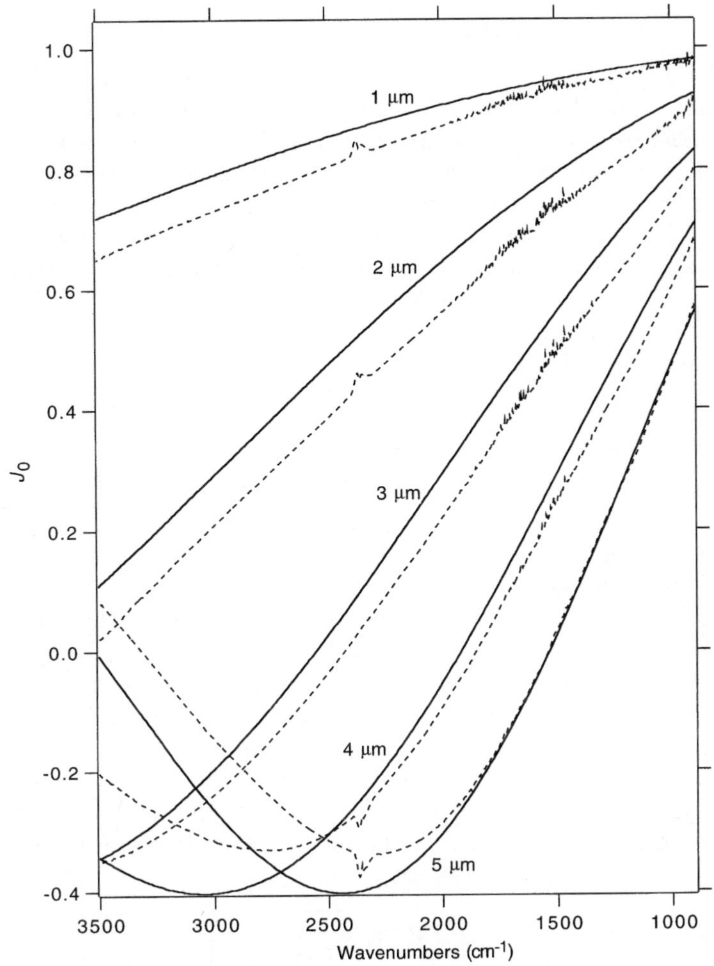

Figure 17 The experimental (dashed lines) and predicted (solid lines) $J_0(\delta^0_{\varphi_i})$ spectra. The solid traces with labels 1, 2, 3, 4 or 5 μm represent the $J_0(\delta^0_{\varphi_i})$ spectra predicted when half-wave retardation is introduced by the PEM at those wavelengths

for dichroism measurements over an ADI (see Section 3.3), the importance of PDI/DPM is apparent.

5 Additional applications

In addition to the demonstrated applications discussed earlier, some other applications that are yet to be explored are summarized in this section.

5.1 Optical rotatory dispersion

When linearly polarized light passes through an optically active sample (consisting of chiral molecules) in an isotropic phase (liquid solutions), the plane of polarized light is rotated, and this angle of rotation is wavelength dependent. A plot of the angle of rotation (ϕ) versus wavelength constitutes the optical rotatory dispersion spectrum. When a isotropic chiral sample is placed before the analyzer, as shown in the configuration in Figure 4(e), and the analyzer is oriented with its axis parallel to the direction of wires of the beamsplitter, the resulting interferogram depends on the sample's optical rotatory dispersion. The cosine Fourier transform of this interferogram is proportional [6] to sin ϕ.

5.2 Raman scattering

Raman scattering experiments with PDI have not yet been reported, but an analysis [11] for conducting the Raman scattering measurements has been reported (see Figure 18). The sample of interest is illuminated with linearly polarized monochromatic laser radiation with its electric vector either perpendicular or parallel to the horizontal scattering plane. The wavelength of this exciting laser beam can be in the near-infrared region (~1 μm), where the wire-grid beamsplitters with 0.25 μm spacing are expected to function satisfactorily. The collimated Raman scattered light serves as the incoming light for PDI. When chiral molecules are illuminated with plane polarized monochromatic light, the 90° scattered Raman light can have linear polarization components $I^{p}(\bar{\nu})$ and $I^{s}(\bar{\nu})$ (where superscripts p and s represent the polarizations parallel and perpendicular to the scattering plane, respectively) as

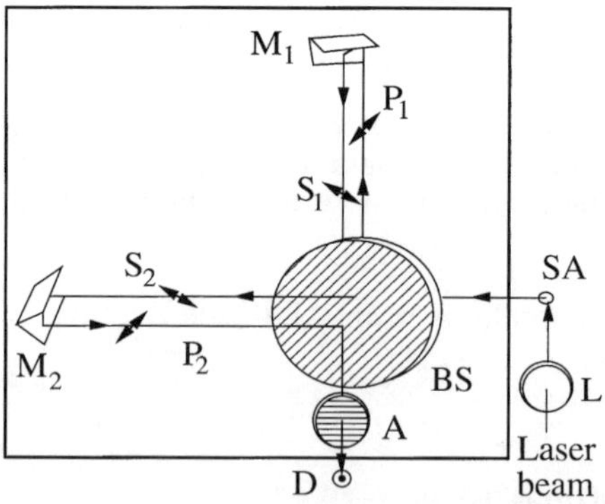

Figure 18 Proposed configuration of a wire-grid beamsplitter-based PDI for Raman spectral measurements in 90° scattering geometry. The definitions for the optical components are as in Figure 2

well as left (L) and right (R) circular polarization components, $I^{L}(\bar{\nu})$ and $I^{R}(\bar{\nu})$. As a result the Raman difference spectrum $[I^{p}(\bar{\nu}_i) - I^{s}(\bar{\nu}_i)]$ can be obtained directly as the cosine Fourier transform of the measured interferogram; the Raman optical activity [25] (ROA) spectrum $[I^{L}(\bar{\nu}_i) - I^{R}(\bar{\nu}_i)]$ can be obtained as the sine Fourier transform of the measured interferogram. The measured ROA represents depolarized ROA when the incident laser polarization is in the horizontal scattering plane and polarized ROA when it is perpendicular to that plane. To measure the depolarization ratios of Raman bands, the incident laser polarization needs to be perpendicular to the scattering plane and a linear polarization analyzer needs to be incorporated in the Raman scattered light. Two measurements with this analyzer oriented parallel and perpendicular to the scattering plane can be used to derive the depolarization ratios.

5.3 Emission spectra

For samples which support differential polarized emission (or fluorescence), the methodology is similar to that given above for Raman measurements.

5.4 Achromatic polarization modulator

As may have been evident in the introductory discussion, a PDI/SPM can be configured to serve as an achromatic polarization modulator [12]. In this case, an output analyzer would not be present and a light source that is either tunable in wavelength or dispersed with a monochromator would have been used. If both mirrors are held fixed at the ZPD point and one of them is jittered, then the desired polarization state for a wavelength of interest can be obtained by adjusting the displacement amplitude of the jittering mirror. Polarization modulation between two orthogonal linear or circular polarizations can be achieved in this manner, at a frequency that is determined by that of the jittering mirror. This concept will be particularly useful for the far-infrared region, where material-based optical components (which have poor transmission) cannot be used as polarization modulators.

6 Future developments

Significant new developments in instrumentation can be expected in the future. Currently available commercial infrared wire-grid polarizers have an extinction ratio of ~100 : 1. In order to measure the polarization difference signals with minimal artifacts, this extinction ratio should be improved significantly, to perhaps 10 000 : 1. When beamsplitters based on such improved wire-grids become available, the quality of measurements will accordingly improve. The experimental demonstration of additional applications, such as measurements of Raman spectra and optical rotatory dispersion, will greatly enhance the interest of practicing scientists in this interferometry.

7 Acknowledgements

I am grateful to NIH (GM29375) and Vanderbilt University for grants awarded to PDI projects. I am indebted to former students, Dr G.-C. Chen and Dr Z. Deng, who participated in the development and applications of PDI; the liquid crystal data presented in Figure 8–11 were measured by Dr Deng. Technical assistance from, and valuable discussions with, Steve Weibel and Mattson Instruments are gratefully acknowledged.

8 References

1. M. Fracon and S. Mallick, *Polarization Interferometers*, Wiley–Interscience, New York (1971).
2. T. Okamoto, S. Kawata and S. Minami, *Appl. Spectrosc.*, **40**, 691 (1986); S. Takahashi, J.S. Ann, S. Asaka and T. Kitagawa, *Appl. Spectrosc.*, **47**, 863 (1993).
3. M.J. Padgett, A.R. Harvey, A. J. Duncan and W. Sibbett, *Appl. Opt.*, **33**, 6035 (1994).
4. D.H. Martin and E. Puplett, *Infrared Phys.*, **10**, 105 (1969).
5. P.R. Griffiths and J.A, de Haseth, *Fourier Transform Infrared Spectrometry*, Wiley, New York (1988).
6 M.J. Dignam and M. D. Baker, *Appl. Spectrosc.*, **35**, 186 (1981).
7. *The Bomem DA3 Based Polarizing Interferometer Accessory*, Technical Bulletin, Bomem, Quebec (1987).
8. Ishida, Y. Ishino, H. Buijs, C. Tripp and M. J. Dignam, *Appl. Spectrosc.*, **41**, 1288 (1987).
9. N. Raghunathan, N.S. Lee, T.B. Freedman, L.A. Nafie, C. Tripp and H. Buijs, *Appl. Spectrosc.*, **44**, 5 (1990).
10. P.L. Polavarapu, *Infrared Phys.*, *28*, 109 (1988).
11. P.L. Polavarapu, *Chem. Phys. Lett.*, *148*, 21 (1988).
12. P.L. Polavarapu, *SPIE Proc.*, *1166*, 472 (1989).
13. (a) P.L. Polavarapu, G.-C. Chen and S. Weibel, *Appl. Spectrosc.*, **48**, 1224 (1994); (b) P.L. Polavarapu, Z. Deng and S. Weibel, *Appl. Spectrosc.*, **50**, 98 (1996).
14. (a) P.L. Polavarapu and Z. Deng, *Appl. Spectrosc.*, **50**, 686 (1996); (b) P.L. Polavarapu and G.-C. Chen, *Appl. Spectrosc.*, **48**, 1410 (1994); (c). S. Weibel and P.L. Polavarapu, *Appl. Spectrosc.*, **44**, 1369 (1990); P.L. Polavarapu, P.G. Quincey and J.R. Birch, *Infrared Phys.*, **30**, 175 (1990).
15. P.L. Polavarapu and Z. Deng, *Appl. Spectrosc.*, **48**, 1562 (1994).
16. P.L. Polavarapu, G.-C. Chen and Z. Deng, *Appl. Spectrosc.*, *48*, 1403 (1994); unpublished results.
17. P.L. Polavarapu, Z. Deng and G.-C. Chen, *Appl. Spectrosc.*, **49**, 229 (1995).
18. P.L. Polavarapu and Z. Deng, *Appl. Spectrosc.*, **50**, 91 (1996).

19. P.L. Polavarapu, *Appl. Spectrosc.*, **51**, 770 (1997).
20. The wire-grid polarizers used in our laboratory have been obtained from (a) Graseby Specac, 301 Commercial Dr., Fairfield, CT 06430; (b) Molectron, 7470 S.W. Bridgeport Rd, Portland, OR 97224; (c) Optometrics USA, Inc., Nemco Way, Ayer, MA 01432; (d) Sciencetech, 45 Meg Drive, Londin, Ontario N6E 2V2, Canada.
21. Poly-IR Fresnel lenses suitable for the far-infrared region were obtained from Fresnel Technology, Inc., 101 West Morningside Dr., Fortworth, TX 76110.
22. (a) L.A. Nafie and D.W. Vidrine, in *Fourier Transform Infrared Spectroscopy*, Eds J.R. Ferraro and L.J. Basile, Academic Press, New York (1982); (b) P.L. Polavarapu, in *Fourier Transform Infrared Spectroscopy*, Eds J.R. Ferraro and L.J. Basile, Eds. Academic Press, New York (1985); (c) T.A. Keiderling, in *Practical Fourier Transform Infrared Spectroscopy*, Eds J.R. Ferraro and K. Krishnan, Academic Press, New York (1990); (d) C. Marcott, A. E. Dowrey and I. Noda, *Appl. Spectrosc.*, **47**, 1324 (1993); (e) D. Tsankov, T. Eggiman and H. Weiser, *Appl. Spectrosc.*, **49**, 132 (1995); (f) R.W. Bormett and S.A. Asher, *Appl. Spectrosc.*, in press.
23. (a) Meadowlark Optics, 7460 Weld Country Rd 1, Longmont, CO 80501; (b) EM Industries, Hawthorn, NY.
24. K. Masutani, H. Sugisawa, A. Yokota, Y. Furukawa and M. Taumi, *Appl. Spectrosc.*, **46**, 560 (1992).
25. L.D. Barron, *Molecular Light Scattering and Optical Activity*, Cambridge University Press, Cambridge (1982); L.A. Nafie, and D. Che, *Adv. Chem. Phys.*, **85**, 105 (1994); P.L. Polavarapu, *J. Phys. Chem.*, **94**, 8106 (1990).

Index